Anthropology and Science

ASA Monographs

ISSN 0066-9679

1. *The Relevance of Models for Social Anthropology*, ed. M. Banton
2. *Political Systems and the Distribution of Power*, ed. M. Banton
3. *Anthropological Approaches to the Study of Religion*, ed. M. Banton
4. *The Social Anthropology of Complex Societies*, ed. M. Banton
5. *The Structural Study of Myth and Totemism*, ed. E. R. Leach
6. *Themes in Economic Anthropology*, ed. R. Firth
7. *History and Social Anthropology*, ed. I. M. Lewis
8. *Socialization: The Approach from Social Anthropology*, ed. P. Mayer
9. *Witchcraft Confessions and Accusations*, ed. M. Douglas
10. *Social Anthropology and Language*, ed. E. Ardener
11. *Rethinking Kinship and Marriage*, ed. R. Needham
12. *Urban Ethnicity*, ed. A. Cohen
13. *Social Anthropology and Medicine*, ed. J. B. Loudon
14. *Social Anthropology and Law*, ed. I. Hamnett
15. *The Anthropology of the Body*, ed. J. Blacking
16. *Regional Cults*, ed. R . P. Werbner
17. *Sex and Age as Principles of Social Differentiation*, ed. J. La Fontaine
18. *Social and Ecological Systems*, ed. P. C. Burnham and R. F. Ellen
19. *Social Anthropology of Work*, ed. S. Wallman
20. *The Structure of Folk Models*, ed. L. Holy and L. Stuchlik
21. *Religious Organization and Religious Experience*, ed. J. Davis
22. *Semantic Anthropology*, ed. D. Parkin
23. *Social Anthropology and Development Policy*, ed. R. Grillo and A. Rew
24. *Reason and Morality*, ed. J. Overing
25. *Anthropology at Home*, ed. A. Jackson
26. *Migrants, Workers, and the Social Order*, ed. J. S. Eades
27. *History and Ethnicity*, ed. E. Tonkin, M. McDonald and M. Chapman
28. *Anthropology and the Riddle of the Sphinx: Paradox and Change in the Life Course*, ed. P. Spencer
29. *Anthropology and Autobiography*, ed. J. Okely and H. Callaway
30. *Contemporary Futures: Perspectives from Social Anthropology*, ed. S. Wallman
31. *Socialism: Ideals, Ideologies and Local Practice*, ed. C. M. Hann
32. *Environmentalism: The View from Anthropology*, ed. K. Milton
33. *Questions of Consciousness*, eds A. P. Cohen and N. Rapport
34. *After Writing Culture: Epistemology and Praxis in Contemporary Anthropology*, eds A. James, A. Dawson and J. Hockey
35. *Ritual, Performance, Media*, ed. F. Hughes-Freeland
36. *The Anthropology of Power*, ed. Angela Cheater
37. *An Anthropology of Indirect Communication*, ed. J. Hendry and C. W. Watson
38. *Elite Cultures*, ed. C. Shore and S. Nugent
39. *Participating in Development*, ed. P. Sillitoe, A. Bicker and J. Pottier
40. *Human Rights in Global Perspective*, ed. R. A. Wilson and J. P. Mitchell
41. *The Qualities of Time*, ed. W. James and D. Mills
42. *Locating the Field: Space, Place and Context in Anthropology,* eds Simon Coleman and Peter Collins

Anthropology and Science

Epistemologies in Practice

Edited by
Jeanette Edwards, Penny Harvey and
Peter Wade

BERG
Oxford • New York

First published in 2007 by
Berg
Editorial offices:
1st Floor, Angel Court, 81 St Clements Street, Oxford, OX4 1AW, UK
175 Fifth Avenue, New York, NY 10010, USA

Berg is the imprint of Oxford International Publishers Ltd.

Library of Congress Cataloging-in-Publication Data

Anthropology and science : epistemologies in practice / edited by
Jeanette Edwards, Penny Harvey, and Peter Wade.
p. cm.—(ASA monographs, ISSN 0066-9679)
Includes bibliographical references and index.
ISBN-13: 978-1-84520-499-0 (cloth)
ISBN-10: 1-84520-499-9 (cloth)
ISBN-13: 978-1-84520-500-3 (pbk.)
ISBN-10: 1-84520-500-6 (pbk.)
1. Anthropology—Philosophy. 2. Science—Philosophy. 3. Science—
Social aspects. 4. Science and civilization. I. Edwards, Jeanette,
1954– II. Harvey, Penelope, 1956– III. Wade, Peter, 1957–

GN33.A447 2007
306.01—dc22

2006038416

British Library Cataloguing-in-Publication Data

A catalogue record for this book is available from the British Library.

ISBN 978 184520 499 0 (Cloth)

ISBN 978 184520 500 3 (Paper)

Typeset by JS Typesetting Ltd, Porthcawl, Mid Glamorgan
Printed in the United Kingdom by Biddles Ltd, King's Lynn

www.bergpublishers.com

Contents

Notes on Contributors

Catherine Alexander teaches in the Anthropology Department of Goldsmiths College, University of London. She works in Turkey, Kazakhstan and Britain and has published on issues of urban governance, privatization, bureaucracy, property relations and the built environment. *Personal States* (2002) explored the various relationships between different levels of the Turkish state-owned enterprise for sugar production; how they were enacted and represented. She is currently writing *Mercurial Cities* (Cornell University Press), an analysis of the social and economic changes in Almaty, Kazakhstan's former capital, since the collapse of the Soviet Union.

Debbora Battaglia is a social anthropologist specializing in expressive culture, cultural identity and science and religion in the public sphere. Most recently she has studied the discourse of human replication in faith-based communities. Her books include *On the Bones of the Serpent: Person, Memory and Mortality on Sabarl Island* (1990), and the anthologies *Rhetorics of Self-Making* (1995), and *E. T. Culture: Anthropology in Outerspaces* (2005). Recipient of major awards from the John Simon Guggenheim Foundation and the National Endowment for the Humanities, she is the author of numerous scholarly articles. She lives in Massachusetts, where she is Professor of Anthropology at Mount Holyoke College.

Eeva Berglund's research has examined environmental politics and the spatial organization of work and culture. She was lecturer at the Department of Anthropology at the University of London's Goldsmiths College from 1998 until 2002. Since then she has been working for the voluntary sector in London as well as writing for academic and non-academic audiences. *Ethnographies of Conservation: Environmentalism and the Distribution of Privilege*, co-edited with David G. Anderson, was published in 2003 by Berghahn.

Alberto Corsín Jiménez is Lecturer in the Anthropology of Organisations at the University of Manchester. He carried out fieldwork among the nitrate-mining communities of the Atacama desert in Chile. He writes mainly on the political and ethical formations of contemporary capitalism, and in particular on the institutionalisation of values and knowledge. He is editor of *The Anthropology* *of Organizations* (2007) and is currently editing a book on *The Anthropology of Well-Being*.

Tony Crook joined St Andrews in 2003 from Edinburgh where he held a British Academy postdoctoral fellowship, and was involved in the Cambridge-Brunel ESRC-funded project 'Property, Transactions and Creations'. His research is based in the Min region of Papua New Guinea and concerns the holographic relations between knowledge-practices, taro gardening, male initiation ritual and, more recently, mining. The British Academy is to publish *Anthropological Knowledge, Secrecy and Bolivip, Papua New Guinea: Exchanging Skin* in 2007.

Jeanette Edwards is Senior Lecturer in Social Anthropology at Manchester University. She has carried out extensive ethnographic fieldwork in the north of England and has published widely on kinship, new reproductive technologies and 'public understanding' of science. She co-ordinated and directed the European-funded project 'Public Understanding of Genetics: a Cross-cultural and Ethnographic Study of the "New Genetics" and Social Identity'. She is co-editor, with Penny Harvey and Peter Wade, of the forthcoming volume, which complements the present one, *Technologized Images, Technologized Bodies: Anthropological Approaches to a New Politics of Vision* (Berghahn).

Sahra Gibbon is a post-doctoral research fellow in the Anthropology Department at University College London. She has a long-standing interest in the social and cultural dimensions of developments in genomics and the dynamic interface between differently constituted sciences and publics, particularly in relation to breast cancer. She is the author of *Breast Cancer Genes and the Gendering of Knowledge* (2007) and the co-editor with Carlos Novas of the forthcoming *Bio-Socialities, Genetics and the Social Sciences* (2007). She has taught a range of courses focusing on the anthropology of medicine, science and technology.

Hugh Gusterson is Professor of Cultural Studies at George Mason University. He is the author of the books *Nuclear Rites* (1996) and *People of the Bomb* (2004), and co-editor of *Why America's Top Pundits Are Wrong* (2005). He writes for popular and academic audiences on nuclear culture, American militarism, and ethics and science.

Penny Harvey is Professor of Social Anthropology at the University of Manchester. She has done fieldwork in Peru, in Spain and in Manchester, UK on the politics of communication, looking at language, exhibitionary forms, information technologies and most recently road building. She is author of *Hybrids of Modernity: Anthropology the Nation State and the Universal Exhibition* (Routledge

1996), and editor of *Technology as Skilled Practice* (1997), special issue of *Social Analysis* 41(1). She is co-editor, with Jeanette Edwards and Peter Wade, of the forthcoming volume, which complements the present one, *Technologized Images, Technologized Bodies: Anthropological Approaches to a New Politics of Vision* (Berghahn).

Amiria Henare is a curator in the Museum of Archaeology and Anthropology at the University of Cambridge, where she also teaches in the Department of Social Anthropology. Her main field sites to date have been in Scotland and New Zealand, where she has studied museums and Maori weaving, and her research interests centre on the role of artefacts, broadly conceived, in anthropological research. Her first book, *Museums, Anthropology and Imperial Exchange*, was published in 2005 by Cambridge University Press.

Nathan Porath received his PhD from Leiden University on shamanic healing and therapy among the Orang Sakai of Riau (Sumatra). He has also carried out anthropological field research among the Meniq hunters and gatherers and ex-hunters and gatherers of South Thailand and on Thai Buddhist/Thai Muslim relations also in the south of Thailand. At present he is focusing on the development of the history of psychiatry in Thailand and Indonesia.

Simon Schaffer is Professor of History of Science in the University of Cambridge. He is the co-author (with Steven Shapin) of *Leviathan and the Air Pump: Hobbes, Boyle and the Experimental Life* (1985). He has published on the social history of experimentation and physical sciences. In 2005 he was awarded a Research Leave Fellowship by the Arts and Humanities Research Council to pursue work on British astronomy and empire in the eighteenth and nineteenth centuries.

Peter Wade is Professor of Social Anthropology at the University of Manchester and works on issues of racial and ethnic identity in Latin America and on practices of racialization in the context of new genomic technologies. His publications include *Race and Ethnicity in Latin America* (1997) and *Race, Nature and Culture: An Anthropological Perspective* (2002). An edited volume, *Race, Ethnicity and Nation: Perspectives from Kinship and Genetics* (Berghahn) is forthcoming. He is also co-editor, with Penny Harvey and Jeanette Edwards, of the forthcoming volume, which complements the present one, *Technologized Images, Technologized Bodies: Anthropological Approaches to a New Politics of Vision* (Berghahn).

Introduction

Epistemologies in Practice

Jeanette Edwards, Penny Harvey and *Peter Wade*

There was some concern expressed during the planning stages of the Fifth Decennial Conference of the Association of Social Anthropologists of the UK and the Commonwealth (the occasion on which the papers collected here were originally presented) that the theme 'anthropology and science' was too narrow and exclusive for an event that should ideally have the ambition of taking stock of current anthropological concerns and interests. We want to emphasize to readers who might not have attended the conference that this volume is neither a collection on the anthropology *of* science nor a revisiting of the once vexed and now distinctly dusty issue of whether anthropology *is* a 'science'.[1] 'Anthropology and science' certainly embraces studies of the institutional production and practice of 'science' but, as we hope you would expect, the ethnographically motivated studies collected here refuse the domain specificity that 'science' quite frequently attempts to invoke. Indeed this refusal of domain specificity, which fascinated and inspired science analyst Bruno Latour, is the hallmark of an anthropological approach that makes it difficult to label these studies as narrow 'anthropologies of science'.[2] Nevertheless they all engage 'science' anthropologically and this brief introduction offers a guide both to what this might mean and to the ways in which it was done.

In some ways this introductory essay is an exercise in intra-disciplinary translation as we seek to present an alternative and far broader conceptualization of 'anthropology's' relation to 'science' than that which provoked the objection referred to above. We hope to convince those anthropologists who might not think that they work on 'anthropology and science' that much of what went on in the 2003 Decennial Conference might be of interest to them. It is also an exercise in contextualization as we aim to present the papers collected here in relation to the wider themes and concerns of the conference which were highly cosmopolitan and addressed topics and interests of significance to a diverse and unbounded anthropology.[3]

Circumscribing Science

The ambitions and pretensions of modern scientific thinking have often constituted a framework for anthropology's attention to human being, providing the motivation to interrogate rather than confirm prevailing scientific orthodoxy. In the field of medical anthropology, for example, there have been sustained critiques of the hegemony of biomedical understandings of the body, disease and well-being (for example, Scheper-Hughes, 1990; Lindenbaum and Lock, 1993) and also a commitment to making biomedicine not only more culturally sensitive but also more effective (for example, Kleinman, 1980; Singer, 1990; Farmer, 1999, 2003; Rapp, 1999). Medical anthropologists have productively unpacked premises that drive cosmopolitan biomedicine through comparison with parallel and alternative understandings of life, death, health and healing (for example, Martin, 1994; Lock, 2002). In development anthropology, too, there have been sharp challenges to rationalist orientations of development practice and to applications of one-size-fits-all models of 'universal knowledge' that promise much but all too often deliver little (see, for example, Richards, 1985; Ferguson, 1990; Lansing, 1991; Hobart, 1993; Gardner and Lewis, 1996; Fairhead and Leach, 2003). Anthropologists working in the field of development have been compelled, like medical anthropologists, to go beyond critique not only to unpack the premises that constitute the notion of 'development' itself (for example, Escobar, 1995), but also to bring to mind parallel epistemologies that have the potential of informing more refined conceptions of 'development' and intervention, adding complexity to simple dichotomies between 'scientific' and 'indigenous/local' knowledge. In addition to such designated 'sub-fields', anthropology has been central to a broader project of dismantling scientific ideas that 'naturalize' racial, gender and sexual differences. It has challenged constructions of 'human nature' that have emerged from sociobiology (Sahlins, 1976) and subsequent forms of neo-Darwinist selectionism (Ingold, 1990).

In advance of their project of critique, anthropologists have drawn fruitfully from the tools of discourse analysis to look at ways in which cultural ideologies infuse scientific objects and techniques. Ludwik Fleck is often cited as a key ancestor in social constructivist accounts of science. As early as 1937, he was writing about how 'pre-scientific' ideas shape what can be seen, despite dominant ideas that science is neutral and draws its findings from empirical reality (Fleck, [1937] 1979). His understanding of the 'pre-scientific ideas' that inform science and its practices and the 'thought collectives' that produce facts have been influential in contemporary social and historical accounts of science. Oudshoorn (1994), for example, draws on Fleck to ask 'what worlds inspired sex hormones?' She shows how particular and prior understandings of gender and sexual difference were taken into the laboratory in the search for sex hormones which in turn circumscribed and moulded what was 'discovered'. Martin (1987) also exemplifies a constructivist approach showing lucidly how dominant cultural ideologies of the time informed scientific representations of gendered bodies. In another classic and often-cited paper she shows how science draws on and reinforces cultural stereotypes of gender differences in its account of human conception, which includes feisty, dynamic sperm and passive, pudding-like ova (Martin, 1991). As these early examples illustrate, anthropologists are well placed to identify and track 'pre-scientific' ideas as well as the ideological underpinnings of contemporary scientific practice in ways that reveal the social, historical and cultural complexity of scientific fact making.

In a different more philosophical trend, anthropological responses to the ways in which 'science' has been circumscribed (both within and beyond anthropology) have generally sought to identify qualities and values in non-modern knowledge systems, which are comparable to those of Western science. This develops earlier anthropological writings, in which 'science' tended to be discussed alongside 'magic' and 'religion' in the contexts of debates on the possibility of diverse rationalities, or of evolutionary stages (Lévy-Bruhl, 1910; Frazer, 1911; Malinowski, 1925; Evans-Pritchard, 1937; Horton, 1967; see also Tambiah, 1990; Lave, 1996; Nader, 1996). Thus experimental attitudes, the search for evidence, coherent argument, openness to refutation, critical thinking, doubt and operational scepticism are all found in life worlds that would not otherwise be described as either 'modern' or 'scientific'.

Lévi-Strauss's depiction of 'the science of the concrete' famously enumerated capacities for abstract conceptualization, systematic categorization and a thirst for objective knowledge among peoples more familiar with mythic thought than with modern science (Lévi-Strauss, 1966). For him the difference between magical and scientific thinking lay in the excesses of the magical imagination rather than in the absence of scientific capacities: 'magical thought can be distinguished from science not so much by any ignorance or contempt of determinism but by a more imperious and uncompromising demand for it which can at the most be regarded as unreasonable and precipitate from the scientific point of view' (Lévi-Strauss, 1981: 11).[4]

In similar vein, Viveiros de Castro (2004) has urged anthropologists to take seriously the theorizing practices of non-Western peoples and not to reduce the scope of the comparative project to specific domains – such as 'science' – that do not necessarily encompass the philosophical understandings of those we study. A relevant example of this approach is his discussion of 'magical thinking' where, drawing on Gell (1988), he argues that 'magic' is not an erroneous or even a culturally specific form of 'science' but rather is the name that anthropologists give to the practices of people who do not recognize the need to divide the universe into moral and physical spheres. The comparison is thus not centred on diverse ways of 'doing science' but rather engages the deeper philosophical foundations of the scientific project and focuses on diverse assumptions concerning the very fabric of the universe and the specificity (or otherwise) of human being.

It is in terms of this wider ontological discussion that ethnographers have pointed to the limited reach of scientific thinking. An anthropological project to understand 'the theories, knowledges, ethical thought and principles of others' (Overing, 1985), to describe and analyse 'how others organize their significative world' (Geertz, 1983), to investigate 'how human beings address quotidian problems, and cope with the finitude, precariousness, and contingency of existence' (Jackson, 1989), reveals the limits of scientific rationality for understanding lived worlds. More recently impelled by what became known as 'the reflexive turn' in the mid-1980s, anthropologists increasingly turned to the study of 'other worlds' closer to home – the theories, knowledges, ethical thought and principles of corporations, institutions, development organizations, and more abstract and mobile conceptual domains such as globalization, consumption, citizenship, technology and science (see for example, Marcus, 1995, Narotzsky, 1997; Ong, 1999; Riles, 2000; Petryna, 2002; Hannerz, 2004; Ong and Collier, 2005).

The dominant notion of modern science as a domain-specific field of enquiry is thus problematic for wider anthropological discussions of knowledge practices. However, within the modern frameworks that impinge on the vast majority of contexts in which anthropologists carry our their ethnographic work, this domain specificity cannot be ignored, but rather calls for ethnographic attention as a cultural ambition in its own right. This abstraction of knowledge from the social conditions of its production, and the parallel capacity to delimit an object of study, was the achievement of the Enlightenment and the integrity and indeed social value of science, at one level, has always rested on its ability to stand outside 'society' and apart from the lived experience of a more complex materiality. The autonomy was also won from the competing claims of other forms of institutionalized knowledge, particularly that of the Church, and of established classical writers but also from the 'ignorance' of the uneducated. The autonomy of science was thus assured in relation to its dedicated expertise, which amounted to an autonomy from the prejudices and ignorance of the non-specialist and the social interests of the powerful, together with a focused and deliberate partiality in relation to the objects of study (and see Harvey, 2006).

Anthropology was (and as our previous examples show, for many continues to be) fully signed up to this view of 'science' as the 'culture of no culture' (Traweek, 1988) as it appeared in the guise of universalizing theory (of its time), deploying scientific theories and methods in its theorization of 'other cultures' – for example, in theories of evolutionism, or more recently of networked global power, or in the search for human universals in domains of learning, cognition, and perception. Furthermore we should not forget the notion explicit in the social sciences, but also integral to all post-Enlightenment science, that modern science, founded on a capacity to abstract knowledge from the social, was nevertheless a deeply social project – holding out the promise of human freedom, emancipation and progress alongside fears of the enslavement to machines, the erasure of human creative agency and the escalation of warfare, disease and environmental degradation (Pfaffenberger, 1992).

More recent interest in modern science as the object of anthropological attention is thus in many respects an extension of a long-standing disciplinary interest in science as a culturally specific (and a highly effective) form of reasoning. Franklin emphasizes this point in her 1995 overview, in which she identifies an increasing interest in the 'cultures of science', the practices and premises of scientific knowledge production, including most particularly its affective, spiritual, political and economic dimensions – in work that is marked by strong engagement with the interdisciplinary fields of science and technology studies (STS) and social studies of knowledge (SSK). Knorr-Cetina (1999) provides a significant exemplification of this field. Looking comparatively at the laboratory practices of high energy physicists and of molecular biologist, she examines how science manufactures knowledge. Her intention is to amplify 'the knowledge mechanisms of contemporary science until they display the smear of technical, social and symbolic dimensions of intricate expert systems' (Knorr-Cetina, 1999: 3). Her critique of the idea that science is homogeneous or monadic and her focus on its disunity is pertinent here, as is her attention to the place of the empirical in scientific investigation and her distinction between object-centred and person-centred relationships in scientific work.

The materiality of science was also at the heart of Traweek's pioneering ethnography of high-energy physics in the US and Japan, which focused on the tension between the timeless knowledge of the 'laws of nature,' the 'negotiable and cumulative beamtimes' of the particle physicists with whom she did fieldwork, and their own limited and intractable lifetimes, their careers (Traweek, 1988). At the heart of this tension lay the detectors, the apparatus that bound these aspects together, registering 'natural' events at subatomic levels and simultaneously serving as a group's means of production of knowledge, the basic tool through which scientists build their careers.

A focus on the agency of objects such as these detectors was at the heart of actor network theory (ANT), which emerged as a central analytical approach in science studies in the 1980s. This was in many senses less surprising or counterintuitive to anthropologists than to some scholars in other disciplines. Anthropologists were already developing ideas about the social lives of things and the biography of 'objects' (see, for example, Appadurai, 1986; Hoskins, 1998). At the same time, ethnography supported the notion that technologies are 'social facts' (Pfaffenberger, 1988, 1992), entailing 'skilled practices' (Harvey, 1997), their efficacy entangled with the social relations through which they acquire the power of transformation and enchantment (Gell, 1988). Actor network theory scholars were influential in moving the debate in new directions. Influenced by ethnographic writing, they set out to decentre the human actor by focusing on the way in which events, resources and actors (both animate and inanimate) are linked in networks, arguing that it is this constellation (network) of things, people, institutions and interests that constitutes scientific knowledge, its facts and its artefacts. Using the language of confluence and assemblage, early actor network theorists such as Callon (1986), Law (1986) and Latour (1987) focused on the way in which knowledge, power and research are co-constructed (and see Jasanoff, 2004). Their interest in the importance of techniques and artefacts to the practices of science led scholars to think more deeply about the entanglements of science and technology, the role of objects in the crafting and embedding of ideas, and the diverse ways in which 'expertise' is attributed.

Most recently the orthodoxy of the 'network' concept has been subjected to examination by anthropologists critical of the ways in which 'networks' have increasingly been taken not simply as abstractions and/or metaphors, but pre-theoretical, self-evident concepts (Moore, 2004). These studies look in detail at how it is that connections are made, sustained, severed or ignored, exploring how people know that they are in a network and what difference it makes to know one is in a network. Ethnographic work on networked socialities (Green et al., 2005; Riles, 2000; Strathern, 1996; Tsing, 2004) seeks to go beyond the assumptions of ANT to examine the more basic anthropological concept of 'the relation' (and see Strathern, 2005). Anthropology and ANT shared a fascination in the ways in which attention to practice reveals networks of collaboration that destabilize those theoretical constructs (often constructs with significant social force) that rest on claims to autonomous reason. However, in general, STS scholars are more interested in showing how specific relations between human and non-human agents are drawn into the business of sustaining the image of an external, objective order of being – the networks that sustain the practices and indeed the very notion of modern science. Anthropologists, by contrast, work ethnographically, looking at how connections are made and unmade between persons, on what terms and with what effects. Here the focus tends to be on the networks through which human beings energize their relationships, and there by explicitly create differences among themselves (Sykes, 2003).

Actor network theory's more philosophical project also lay in revealing the dynamics of what Latour terms 'the modern settlement'. This is the socio-cultural arrangement by which nature and society are kept apart, purified from their hybrid origins and consequently serving the interests of those who, having objectified a world of nature, are then empowered to act upon it, possess, destroy or protect it without having to question the complex social relationships through which 'nature' comes to stand apart in the first place, as facts become facts through the repeated enactment of social interests. Science did not merely depict nature but represented the social interests of the actors enrolled in its production (Latour, 1987, 1993). In relation to our present concerns, the abstraction of science from the social is largely predicated on this idea of 'nature' as autonomous and external to human being and the work of Latour, and of historians of science more generally (for example, the outstanding work of Shapin and Schaffer, 1985; Haraway, 1989, 1991; Kay, 2000), has been important in showing how the twin poles of the modern constitution – nature and politics – are not stable but are constantly collapsing in on each other, generating hybrid forms, and intransigent practices and ideas that refuse the purified state. Again anthropologists were also exploring these issues ethnographically and writing about them in relation to more general fields (such as race, gender, development and health) in which modern Western science clearly had a constitutive effect. The importance of ANT perhaps lay in the provision of a powerful analytic language, and the direct assault made on the self-presentation of much modern science from within traditions of Western philosophy. This work complemented and reinforced much of the ethnographic work which in turn supported the ANT thesis as to the intrinsic indivisibility of 'science' and 'society'.

In *After Nature*, Strathern (1992) alerted us to the way in which ANT was a product of its time. She charts a shift occurring in Euro-American understandings of the role of 'nature' in human affairs, which she dates to the end of the modern epoch. As nature is perceived to be increasingly open to manipulation and intervention it has, Strathern argues, lost its grounding function – no longer acting as the baseline from which 'culture' is elaborated. At the same time, new objects of knowledge are rendered visible by changing scientific techniques and technologies, which in turn feed into the increasing reflexivity and self-awareness that characterizes contemporary modernity. A heightened focus on molecular biology, for example, has resulted not only in the ascendancy of biology and the marriage of biology and biotechnology but also in an acknowledgement by many biologists of uncertainty – contingency is the order of the day. At the same time, the centrality of biotechnology in recent scientific endeavour has presaged novel institutional and funding configurations. The 'new' relationships between, for example, education, research, industry, politics and regulation have resulted in a sense that 'knowledge itself' is being capitalized (Hayden, 2003) – and not only knowledge but 'life itself'. Now nature is not only manipulated but also manufactured and biology and social life merge in the hybrid of biosociality (Rabinow, 1992).

These contemporary contexts of knowledge politics have important implications for our understandings of the relation of science to its publics – the ways in which publics are configured in relation to science as 'users', as 'beneficiaries' or 'opponents', as 'funders', willing or otherwise, as owners of desirable resources and as opinion holders, voters and investors. The following section briefly reviews existing work in this field.

Conflicting Epistemologies: Science and its Publics

Many scientists felt that attack on the narratives of science amounted to an attack on science itself and the 1990s saw a series of vitriolic exchanges, which became known as the 'science wars', in which, in the US at least, jobs and funding were at stake and both scientists and those working under the broad umbrella of science studies felt at best misunderstood and at worst maligned. Some from each 'side' were doubtless engaged on a quest to 'unmask' the other but many did not equate an acknowledgement of the social contingency of scientific production with the undermining of science. Arguments for the social contingency of science clearly had implications for the status of scientific truth claims and for some this was a step too far down the road of contesting the institutional authority of science but many scientists were comfortable with the notion of contingent truths. Of interest here is way in which the 'science wars' again revealed the diversity of science and its varied modes and degrees of institutionalization. Also of interest is the way in which these confrontations revealed a tension between the values of autonomous scientific reasoning and a demand for recognition of the *social* contingency and consequent social responsibility of science. The 'science wars' became an umbrella for a two-pronged refutation of issues that emerged into public debate both through the academic sociology of science, and through an increasing public awareness and concern about the effects, costs and social responsibility of 'big science'. In this way, a new public forum arose for discussion of the critiques of universalism and of institutionalized authority that had been growing since the early and influential publications from within the philosophy of science (Kuhn, 1970; Feyeraband, 1978), taken up and developed in a critical history of science (for example, Latour and Woolgar, 1979; Shapin and Schaffer, 1985; Latour, 1988), in feminist scholarship (for example, Haraway, 1989, 1991, 1997; Harding, 1986, 1991, 1998, 1993; Keller, 1985; Keller and Longino, 1996) and in subaltern responses to ethnocentric accounts of western science (for example Hess, 1995; Prakash, 1999).

Traditionally 'publics' have been problematic for 'science' when science is defined in relation to the search for autonomy from the prejudices of the uneducated, the emotions of the undisciplined and the inherent conservatism of the uncurious. This approach to publics produces what became known as the deficit model of the public understanding of science, a model that clearly reinstates the separation of the scientific and the social and which presumes a need for 'science' to educate, inform and generally promote 'scientific literacy' and for publics to learn well (for example, Claeson et al., 1996; Edwards 2002, 2004). The image is of knowledge that can be transposed intact from expert to non-expert. There have been trenchant critiques of the so-called 'deficit model' particularly from sociology of science perspectives (for example, Wynne, 1991, 1995; Irwin and Wynne, 1996; Nowotny and Taschwer, 1996). Nevertheless, variants of it pop-up in unexpected places and in different guises. Whenever 'lay beliefs' are set in opposition to 'scientific understanding', we have to ask to what effect. At the turn of the nineteenth century, 'primitive thought' was compared to 'scientific thought' and generally found lacking, and science and sorcery were placed on different sides of the civilizing fence. A recent tendency to locate what has been dubbed technoscience in its organizational, political and global settings and traditional knowledge systems with their lay beliefs in local/indigenous settings feeds the current version of a hierarchy of separate and discrete knowledge systems.[5] The new 'primitive' is the scientifically illiterate

Social and historical studies of science have revealed the complex relations and shifting alliances in the social production, transformation and maintenance of science. Its publics, and the connections between them, are numerous and can be found in universities, biotechnology companies, social movements, government, mass media, pharmaceutical industries, and so forth, all of which are linked in unprecedented ways (Rabinow, 1996). One of the arguments in relation to why scientific literacy is necessary is that science (configured as outside the social) abdicates responsibility for the application of its findings to 'society'. Nowotny et al. (2001) identify a shift from attempts to place science more firmly *in* society towards the idea that science should be more accountable to society, and that this is best achieved by bringing *society* into science.[6] It is in this context that we have witnessed the rise of ethical monitoring in Western knowledge institutions. Scientific proposals are increasingly judged by the explicitness with which they identify both the usefulness of their research (to society) and their provisions for ethical monitoring. Research has to be seen to be ethical, and society, via its representatives, 'the public', is to be included at all stages of the project. Scientists are being urged to be cognizant of, to take account of, the wishes and needs of society (Strathern, 2004). In this new regime the public needs not only to understand science, but also to judge it and ideally approve it. But of course, publics are recalcitrant and, trained in the art of exercising individualized consumer rights, will often find ways to do what they want despite legal strictures and expert opinion. Highly informed (often through bespoke information networks) some 'consumers' of science and its applications are able to find their own way. Readers may recall the efforts of Diane Blood who requested that sperm be extracted from her husband who was in a coma. Medical staff complied and his sperm was frozen. But after his death Mrs Blood was not allowed to use the sperm, in part because it was obtained without her husband's informed and written consent.

Eventually she was allowed to carry the frozen gametes of her deceased husband to Belgium where she was inseminated, became pregnant and gave birth to their child. The tenacity of Mrs Blood exemplifies the resistance to domain specificity discussed above: it reveals a biotechnological procedure raising questions to do with kinship, desire, governance (for example, in legislation, a clause concerning the 'right of the child to know its father'), ethics (the limits of 'informed consent'), bureaucracy (the deceased father could not be on the birth certificate of the child conceived with his sperm), 'fertility tourism' and more. In addition, the relative ability of diverse publics to move around the strictures of legal and scientific norms defines another wide field of contemporary ethnographic enquiry. In many cases the possibilities for creating spaces for public action (particularly for public protest) are severely constrained by powerful institutional actors and interests that appropriate the public domain and squeeze out dissenting voices. In other cases, activist groups, campaigners, lobbyists and 'concerned citizens' are successful in changing, or at least halting, for the time being, government policy, sometimes by forging unlikely alliances with those same powerful institutional actors.

Here we have returned, by a somewhat different route, to the co-construction of science and the social – as a variant on another enduring fascination of anthropological enquiry – the place of human being (human intelligence, knowledge, relationships) in relation to their wider environment, the worlds of other beings, agencies, knowledges and so on. Authors in this volume address head-on relationships between science in society but through science that spills over into rationalities of government, bureaucracy and administration; through science that not only refuses domain specificity but flourishes in its multifariousness.

The coexistence of competing epistemologies and practices unites the chapters of this volume and it is to these we now turn as we consider how competing knowledge forms even those ostensibly in collaborative relationships, require a sense of how knowledges are made and maintained, and how they 'gain their shape' in relation to each other (Tsing, 2004: 13).

Epistemologies in Practice

After the dust had settled on the 'science wars', the issue that emerged as significant was how particular objects of science came onto the agenda as significant, what other ideas they are surrounded by and how ideas travel and have effects in the world. Practice approaches to epistemology complicate the narratives of universalism, revealing complexity, ambiguity and a constant spilling over and across domains. As we have seen, anthropologists have worked with these imprecisions in many different ways, exploring dynamics within specific communities of practice, usually in relation to fixed sites, but also by following networked connections across diverse sites. Fine-grained ethnographic studies reveal mutual constructions of global and local knowledges and excavate 'loop-back' mechanisms between the two. They also complicate simple contrasts between lay and expert, local and cosmopolitan, and scientific and indigenous. The studies in this volume combine attention to the material and the discursive, as well as to ideas and practices. The papers presented here are highly diverse in terms of theme and regional focus but are drawn together by a common concern with connections and flows of knowledge: how the connections that constitute knowledge are made and unmade, how certain networks of connections become authoritative, which connections are left implicit and which made explicit and what allows and blocks flows of knowledge.

We start with Schaffer who shows how, in a colonial context, knowledge-generating connections formed between local intermediaries in India and European scientists intent on creating international benchmarks for reliable astronomical observations. But these connections and the reliance of the European scientists on Indian assistants and observers are erased in histories that seek to purvey science as universalist. Science as globally authoritative knowledge is made to 'stick' by erasing these connections. These connections between European and non-European knowledge practitioners are taken up by Henare in her account of the history of connections between Maori and non-Maori anthropologists. She argues that the long-standing links and dialogues between non-Maori and Maori ways of knowing about Maori have been sidelined in post-'writing culture' by non-Maori interests in analysing and deconstructing 'invented traditions' and pondering issues of (strategic) essentialisms in indigenous representations of self. This blinds non-Maori anthropologists to possible alternative ontologies or ways of thinking about being Maori that are radically other and not constrained by notions of invention and the idea of a single reality that they entail. By challenging Western epistemology, Maori anthropologists are pursuing a different agenda from Indian astronomers but the way certain knowledge connections are silenced is a common theme.

The manipulation of connections is addressed by Corsín Jiménez who demonstrates how, in the management of Chilean nitrate mining towns in the 1920 and 1930s, the administrative deployment of knowledge in the pursuit of rational management reveals and hides connections and information and opens and closes channels for the flow of knowledge in ways that allow business to balance 'public service' and private profit to its advantage. The boundary between the public and the private is manipulated through administration. A process such as audit, which appears to be all about transparency and openness to the public gaze, is actually intensely inward looking and self-involved: 'audit is administration gone paranoid rather than public'. Alexander and Berglund both focus on the way knowledge practices are made to work in projects of development and planning and in challenges to those projects. In Almaty, in Kazakhstan, Alexander analyses how Soviet planners produced knowledge of the city as a network of natural and social connections, harmonized by science, for the good of society and its citizens. But the knowledge practices and social networks of central Soviet planners and local city planners were different and came into conflict. Meanwhile the scientific and instrumental nature of national planning was seen by locals as unsociable; they made different connections.

In Finland (discussed by Berglund), certain plausible connections were constructed in the late twentieth century between being a 'new information society' and being a dynamic, innovative and modern nation. This created a counter-image of the older Finland, in which agro-forestry was linked to development, as slow and static. Yet this image silenced the strong connections between agro-forestry science, modernity and national identity that had existed before and that, far from being an imposition by government experts, was a social phenomenon that engaged a collective sense of modern Finnish identity.

Gusterson gives an analysis of how, less obviously than in the Soviet and Finnish cases, scientific knowledge is connected to the political sphere through connections that appear and disappear. He shows that geo-political ways of knowing, which are constituted in the corridors of power in Washington, amid concerns with changing nuclear weapon scenarios, shape and are shaped by the knowledge practices of nuclear lab scientists. A shift in lab science from organizing field tests of real nuclear devices towards conducting computer simulations paralleled the demise of a Cold War nuclear stand-off and a growing concern with controlling all nuclear materials in the context of emerging nuclear 'rogue' states. Changing geo-political futures open and close funding streams that facilitate different scientific ways of knowing about nuclear weapons.

Gibbon traces a similar set of effects, which occur in medical knowledge, this time focusing on the connections between medical expertise on genetic testing for breast cancer and lay knowledge of kinship and inherited conditions. She argues that ideas about family relationships shape, in various ways, both expert and lay ways of knowing about medical genetics, creating a collective endeavour in the work of predictive testing. DNA is not devoid of affect but is strongly imbued with it in the clinical setting. Ideas about obligations and duties between kin influence the way both clinicians and individuals undergoing tests think about how to bring kin into processes of predictive testing but in varied ways. Asking a person's relative to take a test in order to help the predictive process along might be seen by the clinician as offering a useful service to the relative, whereas the person undergoing the test or her relative might regard it as an unwanted and threatening imposition.

Battaglia's essay takes a different approach. She is interested in how the monstrous and the uncontrollable – here in the shape of the Raelians and their claims to have created a cloned human called Eve – can short-circuit the knowledge-production networks of science and create an abyss of non-connection and non-knowledge. The Raelians and their claims are a monstrous example but they act as a figure of the emergent life that media, and especially new media, are predisposed to participate in. Her essay is itself an example of making connections in unpredictable ways.

The spectre of the abyss – the collapse of connection – raises issues of (Western, scientific) epistemology that Crook and Porath both address. Crook takes issue with the figure of the outside observer who can legitimate knowledge. Current anthropological interest in knowledge can construe it as a universal human tool, albeit existing in culturally relative forms, and this is rather similar to notions of a scientific epistemology that can take possession of the facts of the world. Using ethnography from Bolivip in Papua New Guinea to suggest other approaches, Crook suggests that what people are – in terms of 'age, sex, gender, aptitude and respectfulness' – shapes the way they know. The connections of knowers to their social environments and their physical constitutions are important. For Porath, the Cartesian disconnections between mind and body, which we know as the classic mind/body dualism, have been seen as the basis for a scientific posture towards knowledge. This has created a popular opposite image of a non-dualist, non-scientific posture, assumed to be characteristic of many non-Western, or holistic, ontologies. In the process, the possibility of a non-scientific yet dualist mode of knowing – which the Sakai of Sumatra use – has been overshadowed and made invisible.

The papers selected here cannot be considered representative of the conference as whole but rather are exemplary. A sister volume provides another selection of papers from a different but connected perspective: this time from the vantage point of the body and visualizing technologies (Edwards, Harvey and Wade, forthcoming). However to have done justice to the wide range of excellent papers presented at ASA 2003 and to the diversity of thoughtful and provocative panels convened we would have needed either a much longer book or several volumes: a possibility unfortunately curtailed in the present-day political economy of publishing. Nevertheless, we hope you agree that the chapters selected here lay to rest the concerns initially expressed about the possible narrowness of the topic 'Anthropology and Science'.

Acknowledgements

We would like to extend our thanks to: all the convenors of the panels at the ASA Decennial Conference 2003 who, in addition to their notable efforts in arranging an impressive range of panels, assisted us with the initial selection of papers for this volume: Rohan Jackson, an excellent conference organizer; staff in the office of the Department of Social Anthropology; our postgraduate student assistants and the bodies that sponsored the conference: Wenner-Gren Foundation, University of Manchester and the British Academy.

Notes

1. We also want to emphasize to readers who were at the conference that this is a highly selective and selected sample of papers from the roughly 450 that were presented: a selection that does not claim representative status but might fruitfully be read in conjunction with *Technologized Images, Technologized Bodies: Anthropological Approaches to a New Politics of Vision*, edited by Jeanette Edwards, Penny Harvey and Peter Wade (Oxford: Berghahn, forthcoming).
2. See Latour (1993) and, for discussion of this book, Strathern (1996).
3. We are indebted to Marilyn Strathern for her elaboration of the difference between translation and contextualization in her British Academy Isaiah Berlin Lecture, Manchester, March 2006.
4. See also Battaglia this volume, whose analysis of Realian science resonates with Lévi-Strauss's understanding of the idea of 'magical thought' explaining too much, rather than too little.
5. Nader (1996) argues for a mutually informative dialogue between students of technoscience and of ethnoscience. For her, the challenge lies in conceptualizing public understanding of science in ways which neither romanticize 'the people' nor demonize 'the scientist', while at the same keeping a critical eye open to the authoritative nature of a science and technology which has a tendency to colonize and push aside other ways of knowing.
6. Nowotny and colleagues also delimit the new ideal communicative venue – the public space of the agora where ideas can be debated, negotiated and scientists and publics are mutually informed.

Bibliography

Appadurai, A. (ed.) (1986), *The Social Life of Things: Commodities in Cultural Perspective*, Cambridge: Cambridge University Press.

Callon, M. (1986), 'The Sociology of an Actor-network', in M. Callon, J. Law and A. Rip (eds), *Mapping the Dynamics of Science and Technology*, London: Macmillan.

Claeson, B., Martin, E., Richardson, W., Schoch-Spana, M., and Taussig, K. (1996), 'Scientific Literacy, What It Is, Why It's Important and Why Scientists Think We

Don't Have It', in L. Nader (ed.), *Naked Science: Anthropological Inquiry into Boundaries, Power, and Knowledge,* New York: Routledge.

Edwards, J. (2002), 'Taking "Public Understanding" Seriously', *New Genetics and Society*, 21(3): 315–25.

—— (2004), 'Public Understanding of Science: Citizenship, Deficit and Democracy', *Anthropological Journal on European Cultures,* 13: 51–68.

Edwards, J., Harvey, P. and Wade, P. (forthcoming), *Technologized Images, Technologized Bodies: Anthropological Approaches to a New Politics of Vision*, Oxford: Berghahn Books.

Escobar, A. (1995), *Encountering Development: the Making and Unmaking of the Third World*, Princeton, NJ: Princeton University Press.

Evans-Pritchard, E. E. (1937), *Witchcraft, Oracles and Magic among the Azande*, Oxford: Oxford University Press.

Fairhead, J. and Leach, M. (2003), *Science, Society and Power: Environmental Knowledge and Policy in West Africa and the Caribbean,* Cambridge: Cambridge University Press.

Farmer, P. (1999), *Infections and Inequalities: the Modern Plagues,* London: Berkeley: University of California Press.

—— (2003), *Pathologies of Power: Health, Human Rights, and the New War on the Poor*, Berkeley: University of California Press.

Ferguson, J. (1990) *The Anti-Politics Machine: 'Development', Depoliticization and Bureaucratic Power in Lesotho*, Cambridge: Cambridge University Press.

Feyerabend, P. (1991), *Science in a Free Society*, London: NLB Press.

Fleck, L. ([1937] 1979), *Genesis and Development of a Scientific Fact,* Chicago: University of Chicago Press.

Franklin, S. (1995), 'Science as Culture: Culture of Science', *Annual Review of Anthropology*, 24: 163–84.

Frazer, J. G. (1911), The *Golden Bough: A Study in Magic and Religion*, London: Macmillan.

Gardner, K. and Lewis, D. (1996), *Anthropology, Development and the Post-modern Challenge,* London: Pluto Press.

Gell, A. (1988), *Art and Agency: An Anthropological Theory*, Oxford: Clarendon

Geertz, C. (1983), *Local Knowledge*, London: Fontana Press.

Green, S., Knox, H. and Harvey, P. (2005) 'Scales of Place and Networks: an Ethnography of the Imperative to Connect through Information and Communications Technologies', *Current Anthropology*, 46(5): 805–26.

Hannerz, U. (2004), *Foreign News: Exploring the World of Foreign Correspondents*, Chicago: University of Chicago Press.

Haraway, D. J. (1989), *Primate Visions: Gender, Race and Nature in the World of Modern Science*, New York: Routledge.

—— (1991), *Simians, Cyborgs and Women: The Reinvention of Nature*, New York: Routledge.

—— (1997) *Modest_Witness@Second_Millennium.FemaleMan©_Meets_OncoMouse™: Feminism and Technoscience*, London: Routledge.

Harding, S. (1986), *The Science Question in Feminism*, Milton Keynes: Open University Press.

—— (1991), *Whose Science? Whose Knowledge? Thinking from Women's Lives*, Milton Keynes: Open University Press.

—— (ed.) (1993), *The 'Racial' Economy of Science: Toward a Democratic Future*, Bloomington: Indiana University Press.

—— (1998), *Is Science Multicultural? Postcolonialism, Feminisms, and Epistemologies*, Bloomington: Indiana University Press.

Harvey, P. (ed.) (1997), *Technology as Skilled Practice: Approaches from Anthropology, Psychology and History*, special edition of *Social Analysis* 41(1).

—— (2006), 'Arresting Mobility or Locating Expertise: "Globalisation" and the "Knowledge Society"', in M. Lien and M. Melhuus (eds), *Holding Worlds Together*, Oxford: Berghahn Books.

Hayden, C. (2003), *When Nature Goes Public: the Making and Unmaking of Bioprospecting in Mexico*, Princeton: Princeton University Press.

Hess, D. (1995), *Science and Technology in a Multicultural World,* New York: Columbia University Press.

Hobart, M. (1993), *An Anthropological Critique of Development: the Growth of Ignorance*, London: Routledge.

Horton, R. (1967), 'African Traditional Thought and Western Science', *Africa*, 37: 50–71, 155–87.

Hoskins, J. (1998), *Biographical Objects: How Things Tell the Stories of People's Lives*, New York: Routledge.

Irwin, A. and Wynne B. (eds) (1996), *Misunderstanding Science: the Public Reconstruction of Science and Technology*. Cambridge: Cambridge University Press.

Jackson, M. (1989), *Paths Toward a Clearing. Radical Empiricism and Ethnographic Inquiry*, Bloomington: Indiana University Press.

Jasanoff, S. (ed.) (2004), *States of Knowledge: The Co-Production of Science and the Social Order*, London: Routledge.

Kay, L. E. (2000), *Who Wrote the Book of Life? A History of the Genetic Code*, Stanford, CA: Stanford University Press.

Keller, E. F. (1986), *Reflections on Gender and Science*, New Haven: Yale University Press.

Keller, E. F. and Longino, H. E. (eds) (1996), *Feminism and Science*, Oxford: Oxford University Press.

Knorr-Cetina, K. (1999), *Epistemic Cultures: How the Sciences Make Knowledge*, Cambridge, MA: Harvard University Press.

Kuhn, T. S. (1970), *The Structure of Scientific Revolutions,* 2nd edn, Chicago: University of Chicago Press.

Lansing, S. (1991), *Priests and Programmers: Technologies of Power in the Engineered Landscape of Bali*, Princeton: Princeton University Press.

Latour, B. (1987), *Science in Action: How to Follow Scientists and Engineers Through Society*, Cambridge, MA: Harvard University Press.

—— (1988), *The Pasteurisation of France*, Cambridge, MA: Harvard University Press.

—— (1993), *We Have Never Been Modern*, Cambridge, MA: Harvard University Press.

Latour, B. and Woolgar, S. (1996), *Laboratory Life: The Construction of Scientific Facts*, Princeton: Princeton University Press.

Lave, J. (1996), 'The Savagery of the Domestic Mind', in L. Nader (ed.), *Naked Science: Anthropological Inquiry into Boundaries, Power, and Knowledge*, New York: Routledge.

Law, J. (ed.) (1986), *Power, Action and Belief*, London: Routledge & Keegan Paul.

Lévy-Bruhl, L. ([1910] 1979), *How Natives Think*, New York: Arno Press.

Lévi-Strauss, C. (1966), *The Savage Mind*, Chicago: University of Chicago Press.

Lindenbaum, S. and Lock, M. (eds) (1993), *Knowledge, Practice and Power: The Anthropology of Medicine in Everyday Life*, Berkeley: University of California Press.

Lock, M. (2002), *Twice Dead: Organ Transplants and the Reinvention of Death*, Berkeley, CA: University of California Press.

Malinowski, B. ([1925] 1948), *Magic, Science and Religion and Other Essays*, Garden City, New York: Doubleday Anchor.

Marcus, G. (ed.) (1995), *Technoscientific Imaginaries: Conversations, Profiles and Memoirs*, Chicago: University of Chicago Press.

Martin, E. (1987), *The Woman in the Body: A Cultural Analysis of Reproduction*, Boston: Beacon Press.

—— (1991), 'The Egg and the Sperm: How Science has Constructed a Romance Based on Stereotypical Male-female Roles', *Signs*, 16(3): 485–501.

—— (1994), *Flexible Bodies: Tracking Immunity in American Culture from the Days of Polio to the Age of AIDS*, Boston: Beacon Press.

Nader, L. (ed.) (1996), *Naked Science: Anthropological Inquiry into Boundaries, Power, and Knowledge*, New York: Routledge.

Narotzsky, S. (1997), *New Directions in Economic Anthropology*, London: Pluto.

Nowotny, H. and Taschwer, K. (eds) (1996), *The Sociology of the Sciences*, 2 Vols., Cheltenham: Edward Elgar.

Nowotny, H., Scott, P. and Gibbons, M. (2001), *Re-thinking Science: Knowledge and the Public in an Age of Uncertainty*, Oxford: Polity.

Ong, A. (1999), *Flexible Citizenship*, Durham: Duke University Press.

Ong, A. and Collier, S. (eds) (2005), *Global Assemblages: Technology, Politics, and Ethics as Anthropological Problems*, Oxford: Blackwell Publishing.

Overing, J. (ed.) (1985), *Reason and Morality*, London: Tavistock.

Oudshoorn, N. (1994), *Beyond the Natural Body: an Archaeology of Sex Hormones*, London: Routledge.

Petryna, A. (2002), *Life Exposed: Biological Citizens after Chernobyl*, Princeton: Princeton University Press.

Pfaffenberger, B. (1988), 'Fetishised Objects and Humanised Nature: Towards an Anthropology of Technology', *Man* (N. S.) 23: 236–52.

—— (1992), 'Social Anthropology of Technology', *Annual Review of Anthropology*, 21: 491–516.

Prakash, G. (1999), *Another Reason: Science and the Imagination of Modern India*, Princeton: Princeton University Press.

Rabinow, P. (1992), 'Artificiality and Enlightenment: From Sociobiology to Biosociality', in J. Crary and S. Kwinter (eds), *Incorporations*, New York: Zone.

Rapp, R. (1999), *Testing Women, Testing the Fetus*, New York: Routledge.

Richards, P. (1985), *Indigenous Agricultural Revolution*, London: Unwin Hyman.

Riles, A. (2000), *The Network Inside Out*, Ann Arbor: University of Michigan Press.

Sahlins, M. (1976), *Culture and Practical Reason*, Chicago: University of Chicago Press.

Shapin, S. and Schaffer, S. (1985), *Leviathan and the Air-Pump: Hobbes, Boyle and the Experimental Life*, Princeton: Princeton University Press.

Scheper-Hughes, N. (1990), 'Three Propositions for a Critically Applied Medical Anthropology', *Social Science and Medicine*, 30(2): 189–97.

Singer, M. (1990), 'Reinventing Medical Anthropology: Towards a Critical Realignment', *Social Science and Medicine*, 30(2): 179–87.

Strathern, M. (1992), *After Nature: English Kinship in the Late Twentieth Century*, Cambridge: Cambridge University Press.

—— (1996), 'Cutting the network', *Journal of Royal Anthropological Institute* (N. S.) 2: 517–35.

—— (2004), *Commons and Borderland*, Oxford: Sean Kingston Publishing.

—— (2005), *Kinship, Law and the Unexpected*, Cambridge: Cambridge University Press.

Sykes, K. (2003) 'My Aim is True: Postnostalgic Reflections on the Future of Anthropological Science', *American Ethnologist*, 30(1): 156–68.

Traweek, S. (1988), *Beamtimes and Lifetimes: The World of High Energy Physics*, Cambridge, MA: Harvard University Press.

Tambiah, S. J. (1990), *Magic, Science, Religion, and the Scope of Rationality*, Cambridge: Cambridge University Press.

Tsing, A. (2004), *Friction: An Ethnography of Global Connection*, Princeton: Princeton University Press.

Viveiros de Castro, E. (2004), 'Le don et le donné : trois nano-essais sur la parenté et la magi', *Ethnographiques.org*: *Revue en ligne de sciences humaines et sociales*, 6, November 2004. Available online at www.ethnographiques.org.

Wade, P. (2002), *Race, Nature and Culture: an Anthropological Perspective*, London: Pluto Press.

Wynne, B. (1992), 'Misunderstood Misunderstanding: Social Identities and Public Uptake of Science', *Public Understanding of Science*, 1(3): 281–304.

—— (1995), 'Public Understanding of Science', in S. Jasanoff, G. Markle, J. Peterson and T. Pinch (eds), *Handbook of Science and Technology Studies*, London: Sage.

–1–

Astrophysics, Anthropology and Other Imperial Pursuits

Simon Schaffer

Observatories of Modernity in 1900

This book reflects on the relation between anthropology and the sciences. This relation is peculiarly striking in the sites where anthropological and scientific knowledge are made. Ethnographers of science treat laboratories as the chief institutions of modernity (Knorr-Cetina, 1995). Anthropological critique views colonies as modernity's laboratories (Rabinow, 1989; Stoler, 1995). Both projects urge that laboratories in particular, modern knowledge institutions generally, are home to intense control. Laboratory workers exercise a colonial imperium (Latour, 1988) whereas imperial officers exercise experimental control (Osella and Osella, 2000: 9; Philip, 2004: 6). This chapter offers a modest historical reflection on this intriguing claim about colonial projects and the relative standing of their anthropological and scientific enterprises.

It has long been assumed that the laboratory is the dominant institution of modern sciences and forms of controlled experiment the archetypical way of knowing. But museums and observatories hosted at least as significant ways of knowing, standardized classification and ingenious sign-reading (Pickstone, 1994; Sheets-Pyenson, 1988). Because it queries the laboratory's self-evidence and proffers versions of fieldwork as exemplars both for the anthropology of science and the science of anthropology, this chapter tells of a couple of scientific enterprises that have been rarely juxtaposed: field anthropology and observatory astrophysics. Neither sits easily with laboratories' modernity and security. Ethnographic fieldworkers found anthropological laboratories essential yet troubling (Gupta and Ferguson, 1997). Stellar astronomers reckoned astrophysical observatories were merely noisome chemical laboratories (Schaffer, 1995). There have been major changes in models of what counts as a characteristic scientific enterprise. Pattern sciences have shifted from delocalized theoretical reflection on the immediate results of instrumental observation to highly localized engagement with field behaviour.

Several historians of late nineteenth-century sciences have found telling associations between studies of indigenous populations and of solar physics (Bayly, 1996: 261–4; Pang, 2002: 142; Mahias, 1998; Davis, 2001: 216–30). In the cases treated in this chapter, both astrophysical observatories and anthropological surveys were pursued from the hill stations of southern India, emblems and bastions of imperial rule and of its vagaries. Crucial here are the ways of knowing proper to these scientific enterprises. It is no coincidence that in their stories about the relation between these sciences, science historians and anthropologists both attend to the moment of imperialism and modernism around 1900. This is partly because those years saw a pair of methodological programmes: one focused on local ways of reading signs and conjecturing meanings, the other concerned with the manufacture of universal quantitative standards. European scientists began to exploit humbler divination, which produced reliable knowledge of individuals by pursuing inadvertent traces (Ginzburg 1990: 96–123). Long in use among Bengalis, finger-printing was now touted by British officials as a universal method of individuation, while London anthropologists much debated these puzzles of standards and signs (Dirks, 2001: 186–7; Sengoopta, 2003: 101–11; Rabinow 1996: 112–15). But according to another story this was a period of global measures, when large-scale networks of reliable standards were constructed by tying together local values in systems of apparently universal values. New observatories and museums of late nineteenth-century Europe were centres for standards systems. The careers of the new physical sciences of 1900 can scarcely be understood without these systems. The extension of these sciences, especially in colonial regimes, helped make the patchy imperial world systems work. These programmes gave some anthropologists' fieldwork in scientific workplaces an early rationale and now raise questions about histories of astronomers' field sciences. Such a story might help make sense of the relation between sciences' endeavour to manufacture universal standards of measurement and their reliance on situated methods of sign-reading and ingenious conjecture. It will also allow an exploration of the sources of scientific authority and the powers incorporated in these ways of knowing.

Anthropology and Astrophysics in the Hill-stations

Reading signs and establishing standards were both key techniques for the global surveys of 1900 (Kuklick, 1991: 194–9). So Alfred Cort Haddon, as 1903 president of the Anthropological Institute, celebrated 'the intimate connection which is beginning to be felt to exist between anthropological studies in the wider sense and the *practical art* of government' in India (Haddon, 1903: 8). Language of practical art and contest reveals something of the translation and local agency in what has often been seen as effortless subordination. Consider two salient adjacent examples of science during the Raj: ethnographic surveys of the Nilgiri Hills and astronomical surveys from the Palani Hills. Both were run from local colonial headquarters at Madras, the Government Museum and the Government Observatory respectively, then executed in the southern hills. Similar ambiguities applied to the observatories and the survey stations situated in colonial hill stations (Spencer and Thomas, 1948; Kennedy, 1996; Kenny, 1997). In many ways, they were central institutions. There seemingly reliable data might be gathered by the application of unrivalled hardware to pristine phenomena in a state of nature. Yet they were highly marginal, isolated from the centres of their sciences and often deliberately cut off from support and intimacy. This ambiguous geography helps us understand how hill-stations like these worked as observatories of modernity a century ago.

The wooded massif of the Nilgiri Hills has for centuries been the home of the Toda people, pastoralists of sacred buffalos, which provide them with milk and the dairies that form the centres of worship. Europeans, notoriously fascinated with cultural diffusion and isolation, were much tantalized by the Todas (Philip, 2004: 108–10). British power arrived in 1819, accompanied by a French naturalist named Latour. The East India Company began establishing a settlement in the heart of the Toda lands at Ootacamund (now known by its Tamil name, Udhagamandalam). Within five years 'Ooty' (its British sobriquet) had become the chief sanatorium and hill station for British personnel of the Madras Presidency (Walker, 1986: 240–74). Soon large areas of forest were cleared for tea plantations, later for coffee and chinchona. The Todas' image was often defined for imperial consumption through highly standardized studio photography – one 1900 collection of *The Living Races of Mankind* displaying them in 'a sort of tropical Switzerland' (Pinney, 1990: 279–80; Walker, 1997: 107). These Nilgiri institutions and showpieces were subjected to political rituals and metrological regimes of the Raj.

Some 120 km south-east, the British of the Madras Presidency set up another hill station at Kodaikanal in the Palani Hills. Reached via the great pilgrimage shrine at Palani itself, easily cut off in the monsoon, Kodaikanal was founded in 1845 to protect its European residents from heat and cholera. The indigenous population in the Palanis were classified by Raj census-takers as farmers, not pastoralists, so not entitled to the large land grants sanctioned in the Nilgiris. By 1890 there was a European population of almost 2000. Physicians recommended the well-drained soil, the good food and the boating lake (Mitchell, 1972: 102–29). Ooty and Kodai would then become centres for systematic surveys and the exercise of cunning intelligence by all their inhabitants.

The most thorough government survey of the Nilgiris was launched at the century's end as part of an ethnographic census first tried in Bengal in 1884–6 by the Oxford-trained director of ethnography in British India, Herbert Risley (Cohn, 1990: 154–8, 246–7; Dirks, 2001: 49–52, 218–24). Crucial in this mobilization was the notion of demographic 'type' (Edwards, 1990). Techniques dictated by such agencies as the International Statistical Institute in the 1880s were used to turn locally variable accounts of caste into defined things capable of weight and measure (Morrison, 1984). Individuals were supposedly drilled as members of transparent classes visible to a governing power. Risley authored standardized data entry systems for census takers and enthused about 'the great storehouse of social and physical data' accumulated in British India, 'which only need be recorded in order to contribute to the solution of the problems which are being approached in Europe with the aid of material much of which is inferior in quality to the facts readily accessible in India.' Risley's racialist project involved precision correlations between the shape of the human nose, the census categories of caste and achieved social hierarchy (Risley, 1909: 288; Cohn, 1990: 157). He supposed British imperial sciences would weld together a complex of races, types and castes otherwise incapable of self-determination (Raheja, 1999: 128–31; Bates, 1995: 20–23). At the end of 1899, the British Association for the Advancement of Science lobbied the Indian Government to launch anthropometric surveys as part of its census operations, especially of 'Dravidian tribes' and 'jungle races – concerning which our information is very limited'. During 1901 the anthropometric project was launched using up-to-date measuring equipment and 'the only method that is applicable to so vast a population – that of selecting characteristic tribes and measuring a sufficient number of specimens to determine the type' (Risley, 1908).

Scientific metrology continued politics by other means. Risley's nasal indices allegedly gave physical signs of caste and class. It sought to naturalize social facts (Philip, 2004: 111). In the same way, British astrophysicists pursued celestial signs which would allow them to predict the greatest social crises, the catastrophic famines which followed on the failure of the summer monsoons and of the subsequent rice and grain harvests. Holocausts could apparently be naturalized too. Successive government Famine Commissions recorded more than twenty major fatal crises of supply in the half-century after the War of Independence. Astrophysicists were convinced that 'the same Indra and Váyu, the watery atmosphere and the wind, whom the Sanskrit race adored centuries before the commencement of our era, still decide each autumn the fate of the Indian people' (Lockyer and Hunter, 1877: 594). Madras Observatory had been vital in British astronomers' first dim understanding of the rich tradition and sophisticated computational techniques of Tamil astronomy (Raj, 2000). By mastering solar periods, Raj astronomers reckoned they could master what they judged an utterly natural phenomenon. The astrophysicist Norman Lockyer, veteran of several Indian eclipse expeditions, led a campaign to study sunspots and solar spectra to control crises of supply. 'We are destitute,' said British astronomers, 'of the first conditions for a scientific study of the food supply until we reach the period of settled British rule and rain gauges' (Lockyer and Hunter, 1877: 597). None of them conceded that 'settled British rule' helped cause these holocausts. The Parsi intellectual Dadabhai Naoroji, London professor and mathematician, angrily riposted that British administrators were responsible for the deaths of some 10 million Indians in the 1876–9 famines: 'why blame poor Nature when the fault lies at your own door?' (Naoroji, 1901: 216; Davis, 2001: 58). Free-market fatalism sat appallingly well with a search for famines' natural patterns (Davis, 2001: 216–24, 323–6).

These patterns, it was argued, could best be surveyed from new scientific stations set above the sweltering plains. The very highest point in Kodaikanal lies to the west at Nadingapuram Hill where, in 1893, the Government of India decided to build a dedicated observatory to survey the sun (Kochhar, 2002: 24–5). According to John Eliot, head of the Meteorological Survey of the Government of India, 'the situation is almost unique and atmospheric conditions extremely favourable for these observations' (Eliot, 1902). In this astrophysics project, the Cambridge-trained mathematician Eliot played the same role as that discharged in the ethnological census by the high-flying administrator Risley. Both sought natural marks for social realities. Eliot was convinced by Madras astronomers of a cyclical relation between solar activity, monsoons and famines (Arnold, 2000: 134). Innovative hardware and the local expertise of Indian astronomers could analyse the light from sunspots and the whole of the solar surface, yielding signs of the pressure, temperature and movement of the gaseous chromosphere around the sun's core. These signs, in turn, were supposed to unlock the mechanisms of terrestrial weather: 'the cyclic character of the Madras rainfall cycle must henceforth enter into considerations connected with the food supply of the people.' Just as Risley saw the Raj as a 'storehouse' of ethnological material, so the astrophysicists held that 'we have in India a monopoly of the raw material' for solar surveys (Lockyer and Lockyer, 1928: 95–7; Lockyer, 1898: 36). Eliot sought political control over this databank. In the midst of a newly terrible famine which raged from 1898–1900, Eliot had himself named Director-General of Observatories for the entire Raj. Like Risley, he started recruiting local resources and personnel in the south to discharge these scientific programmes (Kumar, 1991: 86–8).

Field Sciences 1895–1910: The Raw Material of Imperial Sciences

Risley's southern Indian representative was Edgar Thurston, a old Etonian medical botanist appointed in 1885 as superintendent of the local 'storehouse of ethnographic data', the Madras Government Museum (Prakash, 1999: 43; Dirks,

2001: 18). Thence he taught policemen the techniques of anthropometry and hoarded material from his fieldtrips (Bourne and Thurston, 1897; Philip, 2004: 128). Museum visitors could see the collection of standard glass eyes Thurston used to classify racial types (Pinney, 1997: 60). The aim was to teach spectators how to see and measure how they behaved in the museum laboratory he built in the 1890s (Thurston, 1898). Thurston first took his tent and equipment to the Nilgiris in 1896. Ootacamund and its surrounding tea estates had already been a site of combined photographic, anthropological and astrophysical work. Twenty years earlier a French expedition surveyed a solar eclipse and made photographs of Toda men and women. This solar data was widely reported in debates on the chromosphere. The Toda images were published in Europe as evidence of the physical parameters of seemingly primitive tribes (Mahias, 1998). So Thurston had good precedents for his new hill-station surveys. Crucial for this work was his Tamil collaborator K. Rangachari (Philip, 2004: 219). Yet for Thurston, the key lesson was a contrast between urbanity and tribalism. He reckoned that 'anthropological research among uneducated and superstitious people who believe in the efficacy of a thread in warding off the evil influence of devils, and are incapable of appreciating that one's motive is quite harmless, requires tact, bribery, coaxing and a large store of patience' (Thurston, 1896: 229). But he also told the school groups who visited his museum how dangerous it was that the Todas were in the very process of being 'civilized' (Thurston, 1896a: 20). Designed allegedly to 'enlighten', surveys' value depended on the absence of 'enlightenment' among the people they studied.

Thurston set up what he called 'an improvised laboratory' in his hill camp and got Todas to visit. Using members of the Malabar police force as standards of physical capacity, he compared Toda strength determined with dynamometers and their skull shape and nasal size measured with Risley's expensive calipers (Risley, 1908; Bates, 1995: 22–3; Philip, 2004: 110–14). When he returned to Ootacamund in 1901, Thurston brought photographs of Todas taken on the first visit ('a source of never-ending interest'), and inadvertently started a competition among the Todas to win the dynamometer game. As in his Madras museum laboratory, so in the hills the fame of Thurston's machine 'spread from *mund* to *mund* and was circulated at funerals' (Thurston, 1896: 144, 229). Skilful in their own preparations for dynamometric showmanship, the Todas also perceived the political motives behind the government's ethnographic survey (Thurston, 1901: 15). Identification and typology were keys to these surveys (Edwards, 2001: 134–5; Pinney, 1997). Thurston projected his huge and chaotic account of *Tribes and Castes of South India*, composed with Rangachari, as a putatively exhaustive tool for imperial government and control. Susan Bayly (1999: 127) has persuasively argued that by turning caste into 'theories of biologically determined race essences', Thurston and his colleagues 'played a critical role in the intellectual history of India and the Empire at large.'

Thurston had his opposite number in the Presidency's astronomical establishment. Madras Observatory was run by the 40-year-old Scotsman, Charles Michie Smith, physics professor at Madras Christian College and director of the local observatory from 1891 (Henderson, 1923). Michie Smith had a thorough grounding in late-Victorian metrology. Just as Madras citizens visited Thurston's museum and his public library, so they used the observatory for their own time standards, weather forecasts and almanacs. These Tamil works were widely used by indigenous experts in celestial prediction. Thus Chintamani Ragoonathachary at Madras Observatory applied his skills as instrumentalist and computer to variable stars. The son of an observatory employee, he became First Assistant, overhauled Tamil almanacs to start what he called a 'native observatory' and lectured on the 1874 Venus transit (Kochhar, 1993: 57). Four Brahmin assistants, led from 1894 by the expert K. V. Sivarama Aiyar, worked at Kodai (Smith, 1900; Kochhar, 1991: 97–9). Michie Smith adapted himself to Eliot's astrophysical programme with efficiency but no enthusiasm (Smith, 1895; Smith, 1898; Smith, 1904). 'The daily presence of abundant raw material in the shape of sunshine must be the first point to be considered in an Indian solar observatory.' Lockyer (1898: 44) 'would recommend that a competent observer be sent from Britain to carry on the work.' Eliot agreed with his friend Lockyer that Michie Smith needed a permanent European colleague at Kodai. They recommended John Evershed, a competent amateur astronomer with considerable experience of eclipse photography in India. By early 1911 Michie Smith had retired and Evershed took over the entire project.

New Sciences 1905–20: The Reorientation of Fieldwork

As a young chemist Evershed had been inspired by Lockyer to build his own spectroscopes, then directed the solar spectroscopy section of the British Astronomical Association, a new organization for amateurs (Stratton, 1957; Evershed, 1955; Maunder, 1899: 65–80). He completely overhauled Kodaikanal's spectroscopic equipment and work patterns. There were many line shifts, typically towards the red end of the spectrum, between solar spectral lines and the equivalent spectra from elements on Earth. Without a reliable standard for spectroscopic measures, the project linking solar and monsoon activity would be compromised (Hentschel, 1997: 206; Hentschel, 1999: 208). Between 1896 and 1910 a range of explanations of solar redshifts were available to astrophysicists – Doppler effects due to motion in line of sight of gases in the solar chromosphere, shifts proportional to the pressure of the luminous gases and anomalous dispersion, which would affect the apparent wavelength of light reaching Earth through the higher layers of the sun's atmosphere (Forbes, 1961; Hentschel, 1997: 228–36). Evershed put his staff to work in teasing these effects apart (Evershed, 1912). He soon commissioned a high-dispersion spectroheliograph at Kodaikanal built by his Madras technicians, the first state-of-the-art astrophysical instrument entirely constructed in India. Then his Brahmin assistants could read off solar features from drawings of projections onto large charts used for recording sunspots and prominences (Evershed, 1911).

The Kodaikanal astronomers decided that there were no pressure effects in the chromosphere. Using Doppler shifts in spectral lines visible from Kodaikanal in conditions cleared by monsoon rain, they claimed in 1909 to see a radial acceleration outwards from the centre of sunspots at deeper levels and an inward motion at their higher levels, a phenomenon soon baptized the 'Evershed effect' (Evershed, 1909). Any motion of this kind would, presumably, vanish at the sun's edge. But Evershed's team saw large redshifts at the solar limb too. Work on the solar limb was aided by a periodic lull in sunspot activity (Evershed, 1913a; Evershed, 1913b). By the outbreak of World War I, when surveys at Kodai were temporarily compromised, Evershed even started to contemplate a possible repulsion exercised by the Earth on incoming sunlight. Evershed confessed the Earth effect was 'very astonishing and rather unexpected'; it was entirely 'contrary to expectation' (Evershed, 1918a; Evershed, 1920b: 155). In the midst of this astonishment Evershed read a paper by the Cambridge astronomer Arthur Eddington, who in 1917 gave his British colleagues a summary of Albert Einstein's novel theory of gravitation, with its apparent implications that starlight should bend under gravitation and that there should be a detectable redshift of about 8 mA in the wavelengths of sunlight reaching Earth (Forbes, 1961: 138–42; Earman and Glymour, 1980b). Kodaikanal set out to make itself a key centre in the pursuit of evidence for the general theory of relativity.

Anthropological surveys in the Nilgiris were also reoriented in these years. Thurston had close links with other biologists in Britain, notably Haddon, who gave Thurston's anthropometry positive reviews and greeted the Madras scientist when Thurston visited Britain on leave in autumn 1895 (Haddon, 1896). Thurston and Haddon stayed close friends throughout the following decade (Thurston, 1904). In spring 1902 Thurston was contacted by another keen British scientist, Haddon's ally the physician and experimental psychologist William Halse Rivers. In his late thirties, Rivers was already a veteran of leading European medical and psychological laboratory programmes and, like Evershed, a customer of Cambridge's scientific instrument workshops. In 1898 Rivers travelled with Haddon and his colleagues to the Torres Straits, taking light tests and clocks, colour and eye tests, dynamographs and electromagnetic machines. Rivers designed new methods of 'field work', a term Haddon now coined explicitly to describe such enterprises as his Torres Strait expedition. In particular, Rivers recognized the collection of kinship data, his so-called 'genealogical method', as a key to unlock social structure in conversations with native informants (Langham, 1981:

69–73, 86–93; Slobodin, 1997: 19–30; Herle and Rouse, 1998: 137–46, 158–78). Rivers contacted Thurston because he wished to try these strategies in southern India, and got funds from the Royal Society and the British Association for the Advancement of Science to pay for his fieldwork. While firmly committed to the value of psychometric tests in the field and better equipped than any previous field worker with high precision laboratory apparatus for field tests of sensory acuity, Rivers was already sceptical of the anthropometric programme that Thurston and his superiors advocated. Instead of ethnology, Rivers sought to launch what he called ethnography, a detailed literary account of the entire social system of a defined group (Kuklick, 1991: 133–49).

Privileged Sites: Relativity and the Predicament of Field Work

The processes through which local scientific performances are compared with other locales are called 'calibration'. Without calibration scientific technique does not easily work anywhere else. Calibration succeeds when the proper community agrees that two different phenomena are in all relevant respects identical (Collins, 1985: 124). The techniques devised in European laboratories, field stations and observatories in the late nineteenth century relied on such calibrations. Reliable machines might make humans reliable delegates, while sensuous experience became entangled with precision measurement. Self-discipline made scientists into self-aware instruments sent out to measure, observe and detect:

> Up till recently ethnology has been an amateur science. The facts on which the science has been based have been collected by people who usually had no scientific training… By means of the genealogical method it is possible to demonstrate the facts of social organization so that they carry conviction to the reader with as much definiteness as is possible in any biological science. (Rivers, 1910: 1–12)

The idea was to use the routines of training to make field experience count elsewhere.

The small size of the seemingly insular Toda community and its status as anthropological exemplar, if not tourist showpiece, made them apt hosts for Rivers' work. Just as Haddon's group in the Torres Straits had relied on local colonial infrastructure of police and administration, so, in the Nilgiris, Rivers made use of Raj government (Rooksby, 1971; Mandelbaum, 1980; Philip, 2004: 109–10). Between June and December 1902, with Thurston in tow, and accompanied as interpreter first by a forest ranger, Albert Urrilla, then by one of the Anglican mission's Tamil catechists, P. Samuel, Rivers worked with the Todas throughout the Nilgiri Hills. Rivers also collaborated with a Toda herdsman, Kodrner, who brought others for psychological testing and acted as guide through the hills. 'My guide at the end of the day would sometimes go for a distance of eight or ten miles and back to arrange for my supply of men for the next day's work.' He recorded the names of every Toda, their genealogies and customs, their dairy cult and their patrilineal clans (Rivers, 1906: 8–10, 14, 19).

The anthropometrist Thurston carefully looked at Todas as though they were anonymous things (Bates, 1995). The anthropologist Rivers instead worked out how named Todas carefully looked at things. He measured his hosts' colour-blindness and their sense of smell, their response to pain and to visual illusions. He commissioned from local studios quite new photographs of Toda dwellings, landscape and customs, he corresponded enthusiastically with botanical and philological experts, he supplied friends among the Todas with medical remedies, spectacles and cash. The Todas themselves, through Samuel, communicated to Rivers their expectations of his return to the Nilgiris, their demands for his photos, and their interest in the possible foreign markets for their fine white cloths (Andrew, 1905). Soon after Rivers left the Nilgiris, his former interpreter, Samuel, wrote to him about arrangements for the Ootacamund photography studio, Wiele and Klein, to make pictures of individual Todas named in the genealogical tables. It has been urged that Rivers innovatively pursued virtual witnessing of Toda ritual by integrating photography with transcription of indigenous language and culture (Hockings, 1992). He helped make images into evidence in Britain and gift objects in the Nilgiris.

By mapping all but a few of the Todas on his charts, Rivers could explain such puzzles as why in a society devoted to marriage between cousins it was to be expected that the same term would be used for father-in-law and for maternal uncle. The method also let him follow some of the complex rituals of Toda life. He rejected picturesque discourse which saw in the Nilgiris a kind of pristine paradise. There were good sociological reasons, Rivers alleged, for Toda behaviour and their choice of such homelands (Rivers, 1906: 466; Rivers, 1903). On his return to his Cambridge college in early 1903, Rivers began a seventy-page paper on 'the senses of the Todas', published in the first number of his own new *British Journal of Psychology* in late 1905, and a 700-page ethnography of Toda society, published the following year (Rivers, 1905). *The Todas*, Rivers (1906: 7) explained, was 'an attempt to apply rigorous methods in the investigation of sociology and religion.' One of the most important implications of Rivers' work was an effective demonstration of the psychic unity of mankind, the claim that differences in psychometric performance were not strongly related to what evolutionists judged the relative cultural development of the subjects on trial. 'Pure sense-acuity is much the same in all races.' Socialization dominated psychological performance. Psychometric testing in the field could plausibly reveal more adequately than any laboratory trials the basic psychological parameters of humanity (Rivers, 1905: 391, 393; Mandelbaum, 1980: 298).

Standardization matters partly because it helps agreement about the virtue of a trial independently of the result of that trial. The puzzle faced both by Rivers and by Evershed was to establish the repute of their collaborators' data when the correct outcome of observations was in question. In spectroscopic astronomy the quality of metrological standards was under threat. Between 1914 and 1923 Evershed and his collaborators had to work hard to make Kodaikanal a reputable observatory of shifts in solar spectral lines. The Indian observatory and meteorology systems offered weighty support. The imperial famine project's status could warrant the quality of the solar spectroscopy work. During 1918, his new chief assistant, Narayana Aiyar, measured redshifts from the sun's centre out to its limb. Kodaikanal's comparatively large shifts could not be marketed uncontroversially to the wider astronomical community. At Mount Wilson, the newly prestigious California observatory that had become astrophysics' metrological centre, Charles St John reported no visible redshifts at the solar centre. Since he could scarcely assail Mount Wilson's equipment, Evershed (1918a: 373) criticized the choice of data: 'because of their faintness, most of the weak lines measured by St John are very difficult and unsatisfactory lines to measure especially in the limb spectra.' St John replied in kind. He alleged that Indian observations were made with inferior equipment and were insufficiently calibrated (Hentschel, 1993b: 151). Evershed (1918a: 373) conceded the point: 'we seem able even now to state, with St John, that our results are distinctly unfavourable to relativity.' But for the Kodaikanal astronomers, like many others, matters shifted decisively after dramatic 1919 reports of light-bending measured during a solar eclipse in the south Atlantic, an announcement stage-managed by Eddington in London at the Royal Astronomical Society. This startling eclipse report in the imperial capital, soon reported by Indian experts in Calcutta, helped make Einstein's general theory of relativity a discovery of cosmological truth (Earman and Glymour, 1980b; Sponsel, 2002; Stanley, 2003). Evershed reorganized observatory work in line with new metropolitan orthodoxy and what he called its 'brilliant confirmation' of general relativity. What had once seemed strange reports of large redshifts from a remote Indian hilltop turned into confirmations of relativistic cosmology. Narayana Aiyar's brilliant observations of large limb shifts, 'taken by themselves, appear favourable to Einstein's theory' (Evershed, 1920b).

Rivers faced very similar problems of replication and reliability. Fieldwork culture had to be transferred elsewhere. This was not easily to be achieved simply by the publication of his results. Rivers tried to enlist the Raj census system for his own purposes. In March 1910, when preparations were being made in Calcutta for the new Indian census, Rivers sent Risley's team schedules of terms of relationship to drill census officers in the genealogical method. Officers ordered several copies of Rivers' work on cousin-marriage throughout the census network (Gait, 1910). Such mobilization did not on its own give demonstrative authority to Rivers' fieldwork but revealed the effort required to make this work count. Rivers also offered, in his lengthy ethnography, what became a talismanic text 'intended to be a sample of scientific method as applied to the collection and recording of ethnographical facts.' Backed with newly standardized methods for anthropological enquirers, authored by Rivers then published by the British Association, *The Todas* was taken as manifesto. 'Details which may seem insignificant or trivial are often of great importance in the comparative study of custom and belief' (Rivers, 1914: vol.1, vi; Urry, 1993: 27–31).

The eminent Cambridge anthropologist James Frazer found Rivers' imagery familiar material for his own solar mythography. He told Rivers in 1906 that 'the book is a model of anthropological research' and that 'every Cambridge anthropologist will feel proud' (Frazer, 1906). The programme functioned as resource for fieldwork, notably that soon pursued in India by Rivers' followers such as Arthur Radcliffe-Brown (in the Andaman islands in 1905) and Charles Seligmann (among the Veddahs in Ceylon in 1906). Seligmann wrote to his London School of Economics colleague Bronislaw Malinowski, most eminent of Rivers' successors, that 'field research in anthropology is what the blood of martyrs is to the Church' (Herle and Rouse, 1998: 165). For Malinowski, in 1910, 'the book on the Todas stands as a model' (Mandelbaum, 1980: 294). In India students of Rivers and Haddon such as G. S. Ghurye at Bombay and L. K. Ananthakrishna Iyer and Sarat Chandra Roy at Calcutta all pursued this kind of work. Such field study, notably that of Sarat Chandra Roy in Chota Nagpur, was often pursued in deliberate contrast to the surveys of Risley's census enumerators (Srinivasan, 1998: 72). So in his review of caste in Indian society, Ghurye insisted that projects such as Risley's census had in fact generated the very 'caste-spirit' they allegedly only reported (Ghurye, 1932: 158). Rivers himself never returned to India, but just before his premature death in 1922 he was planning lectures at the new social anthropology department at Calcutta University. Though eventually excised from the official memory of British social anthropology, for which it was Malinowski who would henceforth act as founding father, Rivers' experiences at Ootacamund and the surrounding hills sought to turn the Nilgiris into an exemplary site for the human sciences and their fortunes elsewhere (Langham, 1981: 179–80; Mandelbaum, 1980: 288–90).

Conclusion: Colonial Science and the Hill-stations

It was hard to make such sites exemplars. One leading British astronomy journal, now proclaiming Einstein as 'the leader of the revolution in science', nevertheless pointed out that unlike Eddington's apparently decisive light-bending observations, in the case of redshifts it was the 'weaker' data, that from India that seemed to support Einstein (Turner, 1921a, 1921b). Evershed protested. His instruments and the local staff were equally splendid, the discovery of new solar physics showed his observatory's virtue and 'the difficulty of deciding for or against the Einstein shift in the Sun lies in the conflicting nature of the evidence itself. It is not a question of weighing the relative accuracy of the measures made [elsewhere] and at Kodaikanal' (Evershed, 1921). In summer 1923 Evershed resigned from the observatory. When in autumn of that year St John at last agreed that all redshifts were consistent with relativistic gravitation, his so-called 'conversion' was lauded as decisive evidence for Einstein's doctrine (Earman and Glymour, 1980a; Hentschel, 1993b: 148–52, 164–5). Kodaikanal's track record was discounted. Evershed portrayed himself as a solitary observer on his hill top observatory. 'Although I had to construct the instruments at Kodai, and did most of the work of exposing the plates and measuring them myself, I managed to get a body of evidence which ought to have been convincing, but does not seem to have made very much impression', he complained. He boasted of the 'extended study' that his team had conducted of solar redshifts; 'yet it seemed that anything from Mount Wilson must necessarily be superior to Kodaikanal' (Evershed, 1923; Turner, 1923). The Kodaikanal programme, reinvigorated after decolonization and now the basis of work at India's principal astrophysics institute at Bangalore, continues to provide precedent for all-India astrophysics. Concern with postcolonial hierarchy stays crucial. Whether in anthropology, astrophysics or meteorology, the dramaturgy of sudden proof at ambitious world-centres can, perversely, swamp painstaking surveys compiled at allegedly more remote out-stations.

Imperial apologists often used stories of empire to describe the role of science in changing the scope of power. Such systems prompt the modern Indian anthropologist Amrit Srinivasan to observe that 'when field-work was colonial, it was understandable that we, as servants of the Raj, acted as modest assistants to the British, sometimes even taking the scholarly place, but only in the data-collecting half of the positivist praxis, leaving science to our betters' (1998: 73). Astrophysical and ethnological surveys manifested this kind of hierarchized politics, because they involved the difficulties of local co-ordination and standardization for the welfare of imperial government. This is why conflicts about status, authority and credibility counted so much. Instead of asking how, in 1900 scientific universality was produced in general, it is better to ask about the conditions at that time that picked out some places and persons as more reliable, secure and indicative. What exactly distinguished, say, Eddington's astronomical trips from Evershed's Kodaikanal, or Rivers' Nilgiris from Malinowski's Trobriands? And, even more importantly, what precisely was it that picked out any surveyor's work in the hills from among the customs of those who lived there?

The hill-station is a perspicuous site at which to pose these questions. It seemed that by constructing their own asylum, governing Europeans could escape the otherwise allegedly inevitable infantilization visited upon all others, thus know and rule them. The highly nuanced categories of hill-station life, with its European elites in hill-top compounds, native bazaars both essential and invisible to their survival, dramatized the gradations of colonial rule (Mitchell, 1972: 18–20). These were not places of utterly incommensurable skills barely in contact; but recruitment to scientific projects involved well-marked forms of hierarchy and dominance. When Evershed sought new assistants at Kodai for 'solar observations', he would stress the 'large amount of measuring work which requires considerable training before the required degree of accuracy is attained'. So he scoured the students at the Madras colleges, explained the difficulties of a transfer from the city to the cool hills, offered subsidized clothing and transport, then sought to integrate the new recruits in the hierarchies of the hill-top team (Mallik, 2000: 667).

European accounts of indigenous incapacity were designed rather to mask the immense reliance of the surveyors on the co-operation and the skills of all others. In these surveys, it has been demonstrated, Indian technicians and experts were essential actors (Pang, 2002: 137–9; Raj, 2002). The same kinds of skilful forms of collaboration and co-operation gradually impressed Rivers in his own fieldwork. Both astrophysicists and anthropologists knew the importance of working at sites already, if unevenly, engrossed by global networks. The predicament of the field sciences around 1900 involved systematic shifts and reconciliations between the values and geography of the transient tented camp, the more permanently established observing base, and the systems of communication and governance on which they utterly relied. This was the pathway in astrophysics too. The general surveys that characterized the enterprise of spectral atlases were displaced by the intensive study of specific groups of carefully chosen emission and absorption lines pursued at Kodaikanal and at its rival and associated observatories. Perhaps most importantly, the construction of these highly specialized spaces, based on cooler hilltops and staffed by supposedly loyal and disciplined personnel in turn demanded the temporary alliance of indigenous and European experts. From the hill-stations new kinds of knowledge emerged that could never quite be dictated by, but entirely depended on, the rules of imperial hierarchy. Imperial policy and overseas distinction have long been taken both as signs and causes of the sciences' universal grip. Here, rather, the focus has been on the exigencies of the field as pattern for a way of knowing. Subtle differences that counted in modern sciences, empire and history, can be illuminated by tracing high sciences' eminences in the messy troubles of lowly life.

Acknowledgements

Work for this chapter was helped by a research leave grant from the Arts and Humanities Research Council. Thanks to Kapil Raj, David Aubin, Charlotte Bigg and Anand Pandian for their generous advice.

Bibliography

Andrew (1903), Letter to W. H. R. Rivers, 13 May, in Cambridge University Library, Haddon MSS envelope 12040.

Arnold, D. (2000), *Science, Technology and Medicine in Colonial India*, Cambridge: Cambridge University Press.

Ayyar, A. A. Narayana (1915), 'On the Displacement at the Sun's Limb of Lines Sensitive to Pressure and Density', *Bulletin of the Kodaikanal Observatory*, 44: 1–9.

Bates, C. (1995), *Race, Caste and Tribe in Central India: The Early Origins of Indian Anthropometry*, Edinburgh: Centre for South Asian Studies, University of Edinburgh.

Bayly, C. A. (1996), *Empire and Information: Intelligence Gathering and Social Communication in India 1780–1870*, Cambridge: Cambridge University Press.

Bayly, S. (1999), *Caste, Society and Politics in India*, Cambridge: Cambridge University Press.

Bourne, A. G. and Thurston, E. (1897), 'Administration Report of the Madras Government Museum' (1896–1897), *Madras Education Proceedings*, no. 553, India Office Records P5278.

Christie, W. (1898), *Report on Indian Observatories with Reference to the Proposed Scheme of Reorganization*, in India Office Records L/E/7/425, Revenue and Statistics Department file 3225/99, fols. 51–59.

Christie, W. and Turner, H. (1898), 'Report on the Expedition to Sahdol', *Monthly Notices of the Royal Astronomical Society*, 58, appendix: 1–21.

Cohn, Bernard (1990), *An Anthropologist Among the Historians and Other Essays*, Delhi: Oxford University Press.

—— (1996), *Colonialism and its Forms of Knowledge: the British in India*, Princeton: Princeton University Press.

Collins, H. M. (1985), *Changing Order: Replication and Induction in Scientific Practice*, London: Sage.

Davis, M. (2001), *Late Victorian Holocausts: el Niño Famines and the Making of the Third World*, London: Verso.

Dirks, N. B. (2001), *Castes of Mind: Colonialism and the Making of Modern India*, Princeton: Princeton UP.

Earman, J. and Glymour, C. (1980a), 'The Gravitational Redshift as a Test of General Relativity: History and Analysis', *Studies in History and Philosophy of Science*, 11: 251–78.

—— (1980b), 'Relativity and Eclipses', *Historical Studies in Physical Science*, 11: 49–85.

Eddington, A. (1917), 'Einstein's Theory of Gravitation', *Monthly Notices of the Royal Astronomical Society*, 77: 377–82.

Edwards, E. (1990), 'The Image as Anthropological Document – Photographic Types: the Pursuit of Method', *Visual Anthropology*, 3: 235–58.

—— (2001), *Raw Histories: Photographs, Anthropology and Museums*, Oxford: Berg.

Eliot, J. (1896), Letter to Denzil Ibbetson, 29 May, in India Office Records L/E/7/425, Revenue and Statistics 3225/99, fols. 91–106.

—— (1902), Letter to William Christie, 23 January, Royal Greenwich Observatory MSS 7/158.

Evershed, J. (1909), 'Radial Movements in Sunspots', *Monthly Notices of the Royal Astronomical Society*, 69: 454–7.

—— (1911), 'Autocollimating Spectroheliograph of the Kodaikanal Observatory', *Monthly Notices of the Royal Astronomical Society*, 71: 719–23.

—— (1912), 'Kodaikanal and Madras Observatories', *Monthly Notices of the Royal Astronomical Society*, 72: 286–7.

—— (1913a), 'Kodaikanal and Madras Observatories', *Monthly Notices of the Royal Astronomical Society*, 73: 258–9.

—— (1913b), 'The Displacement of the Lines of the Solar Spectrum Towards the Red', *The Observatory*, 37: 124–8.

—— (1918a), Letter to Ralph Fowler, 24 September, Royal Astronomical Society Evershed MSS 1 (1) no. 4.

—— (1918b), 'Displacement of Cyanogen Bands in the Solar Spectrum', *The Observatory*, 41: 371–5.

—— (1920a), 'Displacement of the Lines in the Solar Spectrum and Einstein's Prediction', *The Observatory*, 43: 153–7.

—— (1920b), 'Kodaikanal and Madras Observatories', *Monthly Notes of the Royal Astronomical Society*, 80: 395–6.

—— (1923), Letters to Herbert Turner, 19 and 22 December, Royal Astronomical Society Evershed MSS 1 (5) nos. 20–21.

—— (1955), 'Recollections of Seventy Years of Scientific Work', *Vistas in Astronomy*, 1: 33–40.

Forbes, E. (1961), 'A History of the Solar Red Shift Problem', *Annals of Science*, 17: 129–64.

Frazer, J. (1906), Letter to W. H. R. Rivers, Cambridge University Museum of Archaeology and Anthropology, Rivers MSS.

Gait, E. (1910), Letter to W. H. R. Rivers, 25 March, Cambridge University Library Haddon MSS envelope 12031.

Ghurye, G. S. (1932), *Caste and Race in India*, London: Kegan Paul.

Ginzburg, Carlo (1990), *Myths, Emblems, Clues*, London: Hutchinson.

Gupta, A. and Ferguson, J. (eds) (1997), *Anthropological Locations: Boundaries and Grounds of a Field Science*, Berkeley: University of California Press.

Haddon, A. C. (1896), 'Contributions to the Anthropology of British India', *Nature*, 54: 404–5.
—— (1903), 'Report of the Council', *Journal of Anthropological Institute*, 33: 2–8.
Henderson, J. R. (1923), 'C. Michie Smith – obituary', *Proceedings of the Royal Society of Edinburgh*, 43: 253–4.
Hentschel, K. (1993a), 'The Discovery of the Redshift of the Solar Fraunhofer lines by Rowland and Jewell in Baltimore around 1890', *Historical Studies in Physical Science*, 23: 219–77.
—— (1993b), 'The Conversion of St John: a Case Study on the Interplay of Theory and Experiment', *Science in Context*, 6: 137–94.
—— (1997), 'An Unwelcome Discovery: the Pole Effect in the Electric Arc: A Threat to Early Twentieth Century Precision Spectrometry', *Archive for History of Exact Sciences*, 51: 199–271.
—— (1999), 'Photographic Mapping of the Solar Spectrum 1864–1900', *Journal for the History of Astronomy*, 30: 93–119, 201–24.
Herle, A. and Rouse, S. (eds) (1998), *Cambridge and the Torres Strait*, Cambridge: Cambridge UP.
Hockings, P. (1989), 'British Society in the Company, Crown and Congress Eras' in P. Hockings (ed.), *Blue Mountains: the Ethnography and Biogeography of a South Indian Region*, Delhi: Oxford University Press.
—— (1992), 'The Yellow Bough: Rivers's use of Photography in *The Todas*', in Elizabeth Edwards (ed), *Anthropology and Photography 1860–1920*, New Haven: Yale University Press.
Kennedy, D. (1996), *The Magic Mountains: Hill Stations and the British Raj*, Berkeley: University of California Press.
Kenny, J. (1997), 'Claiming the High Ground: Theories of Imperial Authority and the British Hill Station in India', *Political Geography*, 16: 655–73.
Knorr-Cetina, K. (1995), 'Laboratory Studies: the Cultural Approach to the Study of Science', in S. Jasanoff, G. E. Markle, J. C. Peterson, T. J. Pinch (eds), *Handbook of Science and Technology Studies*, Berkeley: Sage.
Kochhar, R. (1991), 'The growth of modern astronomy in India', *Vistas in Astronomy*, 34: 69–105.
—— (1993), 'Science in British India – Indian response', *Current Science*, 64: 53–62.
—— (2002), 'Madras and Kodaikanal Observatories', *Resonance*, 7(8): 16–28.
Kuklick, H. (1991), *The Savage Within: the Social History of British Anthropology 1885–1945*, Cambridge: Cambridge University Press.
Kumar, D. (1995), *Science and the Raj 1857–1905*, New Delhi: Oxford University Press.
Kutzbach, Gisela (1987), 'Concepts of Monsoon Physics in Historical Perspective: the Indian Monsoon', in J. Fein and P. Stephens (eds), *Monsoons*, New York: John Wiley.
Langham, I. (1981), *The Building of British Social Anthropology: W. H. R. Rivers and His Cambridge Disciples in the Development of Kinship Studies, 1898–1931*, Dordrecht: Reidel.

Latour, B. (1988), *The Pasteurization of France*, Cambridge, MA: Harvard University Press.
Lockyer, M. and Lockyer, W. (1928), *Life and Work of Sir Norman Lockyer*, London: Macmillan.
Lockyer, N. (1898), *Report on Indian Observatories and their Organization.* India Office Records L/E/7/425, fols. 24–49.
Lockyer, N. and Hunter, W. W. (1877), 'Sunspots and Famines', *The Nineteenth Century*, 2: 583–602.
Mahias, M. (1998), 'Le soleil noir des Nilgiri: l'astronomie, la photographie et l'anthropologie physique en Inde du Sud', *Gradhiva*, 24: 45–56.
Mallik, D. C. V. (2000), 'K. S. Krishnan and the Kodaikanal Observatory', *Current Science*, 79: 665–8.
Mandelbaum, D. (1980), '*The Todas* in Time Perspective', *Reviews in Anthropology*, 7: 279–302.
Maunder, W. (ed) (1899), *The Indian Eclipse 1898*, London: Hazell, Watson & Viney.
Mitchell, N. (1972), *The Indian Hill-Station: Kodaikanal*, Chicago: University of Chicago Department of Geography.
Morrison, C. (1984), 'Three Styles of Imperial Anthropology: British Officials as Anthropologists of India', *Knowledge and Society*, 5: 141–69.
Naoroji, D. (1901), *Poverty and Un-British Rule in India*, London: Swan Sonnenschein.
Osella, F. and Osella, C. (2000), *Social Mobility in Kerala: Modernity and Identity in Conflict*, London: Pluto.
Pang, A. (2002), *Empire and the Sun: Victorian Solar Eclipse Expeditions*, Stanford: Stanford University Press.
Philip, K. (2004), *Civilizing Natures: Race, Resources and Modernity in Colonial South India*, Piscataway, NJ: Rutgers University Press.
Pickstone, J. (1994), 'Museological Science? The Place of the Analytical/Comparative in Nineteenth Century Science, Technology and Medicine', *History of Science*, 32: 111–38.
Pinney, C. (1990), 'Classification and Fantasy in the Photographic Construction of Caste and Tribe', *Visual Anthropology*, 3: 259–88.
—— (1997), *Camera Indica: the Social Life of Indian Photographs*, London: Reaktion Books.
Prakash, G. (1999), *Another Reason: Science and the Imagination of Modern India*, Princeton: Princeton University Press.
Rabinow, P. (1989), *French Modern: Norms and Forms of the Social Environment*, Chicago: Chicago University Press.
—— (1996), *Essays on the Anthropology of Reason*, Princeton: Princeton University Press.
Raheja, G. G. (1999), 'The Illusion of Consent: Language, Caste and Colonial Rule in India', in P. Pels and O. Salemnik (eds), *Colonial Subjects: Essays on the Practical History of Anthropology*, Ann Arbor: Michigan University Press.

Raj, K. (2000), 'Colonial Encounters and the Forging of New Knowledge and National Identities: Great Britain and India, 1760–1850', *Osiris* 15: 119–34.
—— (2002), 'When Human Travellers become Instruments: the Indo-British Exploration of Central Asia in the Nineteenth Century', in M. Bourguet, C. Licoppe and H. Otto Sibum (eds), *Instruments, Travel and Science: Itineraries of Precision from the Seventeenth to the Twentieth Century*, London: Routledge.
Risley, H. (1908), 'Memorandum on the Ethnographic Survey of India' (23 October) in India Office MSS Eur E.295 /18.
—— (1909), 'Ethnology and Caste', *Imperial Gazetteer of India*, Oxford: Clarendon Press.
Rivers, W. H. R. (1903), 'The Funeral of Sinerani', *The Eagle*, 24: 337–43.
—— (1905), 'Observations on the Senses of the Todas', *British Journal of Psychology*, 1: 321–96.
—— (1906), *The Todas*, London: Macmillan.
—— (1913), 'Report on Anthropological Research Outside America', *Reports Upon the Present Condition and Future Needs of the Science of Anthropology*, Washington: Carnegie Institution Report no. 200, 5–28.
—— (1914), *The History of Melanesian Society*, 2 vols, Cambridge: Cambridge University Press.
Rooksby, R. L. (1971), 'W. H. R. Rivers and the Todas', *South Asia*, 1: 109–21.
Samuel, P. (1903), Letter to W. H. R. Rivers, 20 May, Cambridge University Library Haddon MSS envelope 12040.
Schaffer, S. (1995), 'Where Experiments End: Table Top Trials in Victorian Astronomy', in J. Buchwald (ed.), *Scientific Practice*, Chicago: Chicago University Press.
Sengoopta, C. (2003), *Imprint of the Raj: How Fingerprinting was Born in Colonial India*, London: Macmillan.
Sheets-Pyenson, S. (1988), *Cathedrals of Science: the Development of Colonial Natural History Museums during the Late Nineteenth Century*, Kingston: McGill-Queen's University Press.
Slobodin, R. (1997), *W. H. R. Rivers*, revised edn, Stroud: Sutton.
Smith, C. M. (1895), Letter to William Christie, 25 April, Royal Greenwich Obvservatory MSS 7 / 158.
—— (1898), Letters to William Christie, 24 March and 12 May, Royal Greenwich Observatory MSS 7 / 158.
—— (1900), *Report on the Kodaikanal and Madras Observatories 1899–1900*, Madras: Madras Observatory.
—— (1904), 'Kodaikanal Observatory', *Monthly Notices of the Royal Astronomical Society*, 64: 333–34.
Spencer, J. E. and Thomas, W. L. (1948), 'The Hill Stations and Summer Resorts of the Orient', *Geographical Review*, 38: 637–51.
Sponsel, A. (2002), 'Constructing a Revolution in Science', *British Journal for the History of Science*, 35: 439–68.
Srinivasan, Amriit (1998), 'The Subject in Fieldwork', in M. Thapan (ed.), *Anthropological Journeys: Reflections on Fieldwork*, Hyderabad: Orient Longman.

Stanley, M. (2003), 'An Expedition to Heal the Wounds of War', *Isis*, 94: 57–89.

Stoler, A. L. (1995), *Race and the Education of Desire: Foucault's History of Sexuality and the Colonial Order of Things*, Durham NC: Duke University Press.

Stratton, F. J. M. (1957), 'John Evershed 1864–1956', *Biographical Memoirs of Fellows of the Royal Society*, 3: 40–51.

Thurston, E. (1895), 'The Madras Museum as an Aid to General and Technical Education' (31 October), in 'Administration Report of the Madras Government Museum', *Madras Education Proceedings* (1896) no. 454, India Office Records P5043, appendix E.

—— (1896), 'Anthropology of the Todas and Kotas of the Nilgiri Hills', *Bulletin of the Madras Government Museum*, 1(4): 141–229.

—— (1898), 'Anthropology in Madras', *Nature*, 58: 82–84.

—— (1901), 'Todas of the Nilgiris', *Bulletin of the Madras Government Museum*, 4(1): 1–21.

—— (1904), Letter to A. C. Haddon, 18 May, Cambridge University Library, Haddon MSS envelope 3.

—— (1905), Letter to W. H. R. Rivers, 9 August, in Cambridge University Library Haddon MSS envelope 12040.

Turner, H. (1921a), 'The Relativity Shift in the Solar Spectrum', *The Observatory*, 44: 159–60, 243–5.

—— (1921b), 'The Einstein Tower', *The Observatory*, 44: 373.

—— (1923), Letter to J. Evershed, 31 December, Royal Astronomical Society Evershed MSS 1 (5) nos. 22–3.

Urry, J. (1993), *Before Social Anthropology: Essays on the History of British Anthropology*, Chur: Harwood Academic.

Walker, A. (1986), *The Toda of South India: A New Look*, Delhi: Hindustan Publishing Corporation.

—— (1997), 'The Western Romance with the Todas', in P. Hockings (ed.), *Blue Mountains Revisited: Cultural Studies on the Nilgiri Hills*, Delhi: Oxford University Press.

–2–

Industry Going Public

Rethinking Knowledge and Administration

Alberto Corsín Jiménez

Knowledge is currently undergoing some remarkable institutional re-locations. Universities are re-organizing themselves in inter-disciplinary schools (cf. Becher and Trowler, 2001); they are also signing up collaborative knowledge-transfer agreements with industry (Etzkowitz et al., 1998). Social scientists are sitting in on hospitals' ethical committees and are invited as observers and co-participants in government, industrial and scientific task forces. Anthropologists, to bring the point home, are now appointed to lectureships in the anthropology of design engineering or the anthropology of organizations; they are employed as professional industrial ethnographers and are much sought-after consultants in the advertising and marketing industries (see, for example, Gellner and Hirsch, 2001). Traditional sites of knowledge-production, like laboratories, academic departments and research centres, are opening up to new social partnerships and alliances (cf. Slaughter and Rhoades, 2004). Knowledge, it would seem, is going social.

Helga Nowotny and her associates tell us that modes of producing scientific knowledge have recently undergone a Kuhnian paradigmatic transformation: science's unilateral engagement with society has turned around and the validation of knowledge as scientifically robust is no longer a matter for scientists to resolve on their own (Nowotny et al., 2001). It has become instead a larger social agenda. Today society decides what makes good science. The question of robustness is therefore crucial (Strathern, 2005). Science and society need new terms of association: new public forums (Nowotny et al. call them 'agora') where their stakes in knowledge are allowed to co-evolve: science and society as one intellectual project. The agenda comes with certain requisites for building social capital. We need, for instance, to trust our scientists; our scientists also need to be responsible (O'Neill, 2002a; O'Neill, 2002b). Trust and responsibility emerge thus as society's new idioms of self-exteriorization: what society comes to look like and how it comes to 'know' itself, in the twenty-first century. Trust and responsibility function as a shorthand for social robustness, which is in turn an index of publicness. In this scenario, science's new institutionalization fares as society's public knowledge.

When transposed to the world of institutions, trust and responsibility look like audits. Institutions become accountable to society and hence robust, through the operation of audit. Audits turn an institution inside out for the public (Power, 1994). Now, despite its obvious linkages to the new contemporary assemblages of science and knowledge, it is intriguing how little attention the question of institutional robustness has drawn from the anthropology of science, or social studies of science at large. It is even more remarkable how little attention industry itself, as the premier site of science's 'public-ization', has attracted from scholars (cf. Miller and O'Leary, 1994). (I use the term 'industry' to capture a particular form of the institutionalization of scientific, research, academic knowledge.) What does the encounter between science and society look like when making its appearance as a protégé of industry? Why is the industrial invisible in our accounts of the flow of (scientific) knowledge? And what of industry itself: how does industry 'go public'?

In this chapter I take to the task of answering some of these questions by drawing an analytical analogy between the institutional publics of science and corporations (industry). Corporate welfare programmes make for an interesting comparison with science and society initiatives because the politics of 'public-ization' permeate both in similar ways. The question of delineating 'publics' emerges as a central concern in both agendas, as does the institutional reorganization of knowledge, which on both accounts is shuffled around between the interstices of the state and the corporate sector in novel ways. Another point of comparison appears in the form of an omission: studies of corporate welfare programmes rely copiously on information about the bureaucratic and administrative workings of corporations. Administration, on the other hand, is conspicuously absent from the social study of scientific practices.

My aim in this chapter is to provide an anthropological outline to some of the questions arising from such an analytical comparison. I turn the current concern with Science and Society programmes inside out in order to map out the contours of a hypothetical inquiry: what publics would an Industry and Society programme make appear (cf. Whyte, 1971/[1946])? Indeed, what would a so-called 'public understanding of industry' tell us about knowledge and its social constituencies? Where, in sum, would an institutional anthropology of knowledge take us?

A tentative and provisional answer has to do with administration.[1] Administration re-vertebrates knowledge at the institutional level and carries it forward. Administration is knowledge's first still point: the place where knowledge settles down temporarily and from where it initiates its institutional travails. Administration is, of course, also central to the operation of audit. It is the point where the organization goes public for society. Administration, in sum, is central to the scaling-out of industry (science) into society, a major player in the realignment of the playing field of corporate and public interests. An anthropology of administration provides a privileged location from where to analyse the effective interaction of knowledge, publicness and institutional sociality. The place where knowledge becomes productively public.

Public Epidemiology

Epidemiology moves in and out of society in surprising ways. In his study of pharmaceutical marketing practices in Argentina, Andrew Lakoff cites a psychiatrist, editor of a leading Argentinean journal of psychiatry, to this same effect. Lakoff's informant is commenting derisively about the shift to neuroscientific explanations in American psychiatry. He links the latter to globalization and American imperialism, and observes how such tendencies 'in the same way that they open the market to foreign products and liquidate the state, they liquidate the forms of hospital care, the training criteria, training institutions, and the public university as the centre of knowledge-production' (Lakoff, 2005: 205).

Lakoff's informant is rehearsing arguments and ideas well known to social students of science and anthropology. Science, he remarks cynically, is caught up in fuzzy political webs, where the institutions of the state, welfare and scholarship stand in ambiguous mutual tension and often undergo profound reorientations. As it turned out, however, the cynic himself was eventually caught unaware by the cynical machine. In the late 1990s, Gador, a leading pharmaceutical company, launched a marketing campaign that revolved around anxieties about globalization. The tone and approach of the campaign was well received by the Argentinean psychiatric establishment, more aligned to a type of social psychiatry that is aware and critical of power relations. This was followed up by a second campaign, which sharpened the theme of globalization by refocusing on the symptomatology of vulnerability. The campaign was launched by a symposium on stress, anxiety and depression and was convened by, amongst others, Lakoff's informant.

Gador's bold marketing campaign pre-empted the psychiatrists' critical discourse. They turned marketing into an instrument for critical diagnosis or, as Lakoff puts it, a framework that 'shows awareness of the ethos of the market' (Lakoff, 2005: 206). If the ethos is critical then marketing espouses a critical ideology too. During the late 1990s' Argentinean economic crisis, market strategists tapped into the social epidemiological model favoured by the local psychiatric establishment, using the crisis as the basis for promoting the medicalization of social suffering. Lakoff observes that despite the absence of robust evidence to support the claim, the media, the marketers and the pharmaceutical industry all agreed that 'the economic crisis was driving up psychopharmaceutical sales'

(Lakoff, 2005: 209). Lack of and concern over the permanence of jobs fuelled, they argued, the sale of antidepressants.

Marketers' 'knowledge' of the market helps them invent the market as a social object. More to the point, their technologies of knowledge in fact allow them to re-calibrate the market as a 'public' object. The data on territorial and prescription sales that pharmaceutical audit firms sell to the industry, for example, allow the latter to generate models of the market where this figures as a 'kind of living entity', something that, in the words of one of Lakoff's informants, 'evolves', 'shrinks', and 'grows' (Lakoff, 2005: 201). The ups and downs of the market are used to re-interpret society's mental health state, as when the economic crisis was invoked to explain the rise in the sale of anti-depressants – market sales read as symptoms of public health.

'Public' is not a word that Lakoff uses often.[2] (Although I note here in passing that 'public' is the term used by his informant, the journal's editor, to describe universities' ownership of knowledge.) He does, however, describe the marketers' knowledge as 'private': knowledge that takes a proprietary form, owned by and sold to corporations. This is information laboriously collected by sales representatives and market analysts, extracted from the 'public domain' (Lakoff, 2005: 211) and commodified for private (corporate) use. The institutional movement of social knowledge in and out of different transactional registers works to redefine the transactional spaces themselves: data on prescription sales is turned into proprietorial knowledge; sales data turned into media scares about national mental health; doctors' ideological stances turned into marketing campaigns. The new institutional routes that knowledge takes to gain hold over its own sense of robustness thus points to the way in which private knowledge is in fact gaining purchase over public definitions. We saw an example of this above: how market indicators were being employed to measure and estimate the state of public health – robust knowledge thus redefined privately as public capital (or lack thereof). The transactions and travelling of knowledge thus hold out the public as an encompassing derivative, the socialized remainder through which private relations are ethicalized. In this context, the public emerges as the social arena that legitimates the market. So long as there is a 'public' to which the market can be held accountable to, market relations remain socially robust (ethical) (Strathern, 2005).

The question of how the public is constituted in action is in fact the focus of Cori Hayden's study of bioprospecting agreements in Mexico (Hayden, 2004). Here knowledge of and claims to natural resources are expressed in a variety of idioms, articulated in terms that institutionalize calls on behalf of the 'national patrimony', communal and indigenous traditional knowledge, market entitlements, corporate intellectual property rights, even the practice of 'culture-free' bioscientific research. The 'public sphere', Hayden writes, 'is intimately tied not just to depictions of the properties in nature, but to highly charged questions about potential property in nature.' (Hayden, 2004: 133) She describes, for instance, the project of a leading UNAM (Mexico National University) ethnobotanical team working on analysing traditional herbal medicines. The project is part of the US government's International Cooperative Biodiversity Groups (ICBG) program, an example of the kind of multilateral agreements that came in the wake of the 1992 UN Convention on Biological Diversity (CBD). The CBD was signed to provide a framework for regulating north-south traffic in biological resources and placed a strong premium on benefit sharing and collaboration with source communities. Holding a rather maverick position within the ICBG framework, the UNAM team has for years been carrying out its prospections in urban marketplaces, purchasing and collecting plants from market vendors. In this context, the team has redefined the 'market' and the information therein contained as part of the national 'public' domain. As one of the researchers put it to Hayden: 'It doesn't make sense to treat vendors as sources of knowledge who deserve compensation: when you buy the plants, you buy the information, and no further obligations are involved' (Hayden, 2004: 125). Hayden hastens to add that the team's aim is not to refuse benefits to source communities, but that they have instead enrolled local participants in alternative collaborative alliances (with NGOs, with artisans' cooperatives, with groups of traditional healers) that circumvent the 'logics of territorially-bounded, property-like claims to authorship and dominion' (Hayden, 2004: 127). We see science here, then, opening up its own political space of action, a conduit for the travelling of both research and political knowledge.

Hayden and Lakoff's ethnographies of science in action make for an interesting comparison. They both show how science swims in and out of politics in different registers, and how the political itself is redefined in accordance with such navigations. They also show that the politics of socializing science – of what Hayden calls the 'public-ization' of knowledge (Hayden, 2003) – is not simply, nor only, a situational battle in the staking out of political claims and assets. Politics is played out in different registers from different institutional heights, and there is a different remit and purchase to, say, how the pharmaceutical industry defines its market, to how a group of psychiatrists define welfare or well-being. The same applies to bioprospecting: the ICGM's idea of a 'market' as mediated by, and inducive of, benefit-sharing with communities does not translate neatly onto the UNAM's ethnobotanists' use of urban markets, where herbal knowledge has no authors and flows through truncated, short-term depositions of information. Not all markets emplace, and are emplaced within, the same public. We may redescribe this by saying that both Lakoff and Hayden's analyses point to the different ways in which (scientific, technical) knowledge appears 'public' in and to society: how and where society recognises 'public knowledge' as its own. For not all trades or industries 'go public' in the same way. Some, like pharmaceutical firms, 'industrialize' society to keep their own self-images and self-projections stable; others, like the critical psychiatrists vision of imperialism, 'de-industrialize' society, slicing up the social in ways that make it easier to talk of redistributive justice and public ownership.

The question of where does knowledge start to become productively public, then, is not trivial. And here I would like to suggest that hidden yet animating both Lakoff and Hayden's ethnographies – sometimes more, sometimes less explicitly so – is the presence of industry. Here, in industry, is where the relational meets the institutional: where sociality holds together in the form of and for a public. In this context, industry remains latent to both Lakoff and Hayden's analyses and folds unto itself the poignant politico-economic predicament that underpins both scenarios. For industry is 'political' science in action. Technology, politics and ethics come together as a powerful institutional assemblage in industry and it is this assemblage that opens up the political spaces wherein different 'publics' take shape and become meaningful (cf. Ong and Collier, 2005). There is a further twist to this, however. For if science goes political and public in industry (if industry is science's public face), the question remains as to how industry itself goes public; that is, what in industry remains stable and recognizable as industry, when industry goes social. How, in other words, does industry fare institutionally in recognition spaces other than its own? The answer, I venture to say, is administration.

Administration

Administration is industry's public. There are two ways in which this can be said to be so. One is through the operation of the institution of audit, a topic of much recent scholarly coverage (for example, Power, 1994; Shore and Wright, 1999; Strathern, 2000). Audit makes certain organizational processes explicit to themselves. It promotes organizational self-consciousness and, in this sense, it is an example of public accountability. Audit, however, evinces an administration worried about its own administrative process. It is therefore administration gone paranoid rather than public.

The second mode in which the public figures in administration has to do with administration as a process rather than an object of self-analysis. Before the days when audit worked as a surrogate for proper administration there was a time when administration was audit: where the administrative process itself was conscious of its public or social dimension. Administration, then, not as the public to industry but as industry's (self-) public. This is of course what bureaucracy itself meant for Weber: a system of administration of proportional forms; proportional, that is, to the size and workings of rationalizing modernity (Weber, 1946). Bureaucracy was to industrialism as society was to economy: a social size for a social process.

The question of proportionality, of the sizes of sociality, has often gone missing in critical reviews of the work of bureaucracy. Yet it is crucial. Focus on the political economy of officialdom, or on its statist or managerial ethos, has often led scholars to lose sight of what keeps the social life of bureaucracy in place, acting as the conduit to the circulation and eventual institutionalization of its knowledge flows: administration. Administration is the social size of a certain social process, namely, institutionalized organization. It is different from management and different from bureaucracy because it pervades both yet it is not reducible to either. Analyses of the taxonomic (Douglas, 1986; Handelman, 1981), praxo-symbolic (Herzfeld, 1992) or power-laden (Heyman, 1995) effects of bureaucracy do not always make explicit that whatever bureaucracy does, as a system of knowledge it always does so through administration. (It does so in other ways too, hence the relevance of these analyses, but administration is always already there.) Administration is an industry's working notion of a social relationship: the shape that sociality takes when cut out in the format of institutionalized organization. Administration, in sum, re-sizes the structures of industry, both inside and outside industry, so that whatever industry goes on to become it will always retain, at whatever scale, a sense of self-similarity.[3] Administration may appear in different forms: meetings (Schwartzman, 1987), projects (Lundin and Söderholm, 1998), evaluation committees (Brenneis, 1994), interdisciplinary teams (Strathern, 2004b), audits (Miller, 2003) – yet it is the one structure that will always be mobilized to keep stable a sense of institutional robustness (Strathern, 2005: 467). No administration, no sense of institutional self.

Public and Institutional Robustness

On 21 June 1928, the office of the General Manager of the Chilean headquarters of the Lautaro Nitrate Company in Santiago, sent out a letter to Francisco Simunovic, *administrador* (manager) of the Santa Luisa nitrate refinery in the northern Chilean province of Taltal. The letter acknowledges receipt of a previous correspondence and orders Mr Simunovic to 'bring the matter to a close'. The matter in question is a reported indictment against Carlos Miranda, the refinery's local school teacher, who had opposed and prevented the raising of the national flag on 21 May, the day of the commemoration of the naval battle of Iquique, a Chilean national holiday. In the letter, Mr Simunovic is prompted to travel to Taltal to meet the local governor and persuade him to drop all charges against Mr Miranda. Mr Simunovic is also alerted to the fact that the office of the general manager will shortly be getting in touch with the provincial governor, based in Antofagasta, to make sure that, should the matter reach his hands, he will not pass it on to the Home Office. The letter concludes with an observation of a general kind, which also explains why the firm reacted with such urgency to the whole affair: 'A foreign corporation cannot be exposed to the accusation of not commemorating the country's national holidays nor the larger patriotic sentiment of its workers and school children.'[4]

The letter makes for an extraordinary exemplar. It locates itself firmly in a system of administration (acknowledging previous correspondences); it carries forward a set of orders and mandates that map out the corporation's bureaucratic powers and remit (from headquarters to local nitrate refineries; from municipal authorities to the Home Office) and it shows institutional sensibility of public issues (national history and patriotic sentiment). In fact, the letter shows that the corporation's awareness of its own institutional robustness depends on it taking charge of the public, seriously. Only as a public institution is the corporation robust as a capitalist organization.

The letter makes for an extraordinary exemplar in a programme of industrial 'public-ization'. It is part of the history of the Chilean saltpetre mining industry in the years of its 'industrialization', c.1917–30, when, fearing the crowding-out of Chilean saltpetre by German synthetic nitrates (as indeed eventually happened), local entrepreneurs and British and American capitalists pushed for the technological upgrade and corporatization of the industry. Corporate welfarism, nationalist rhetoric, labour and gender politics, technoscience, and turn-of-the-century aspirations for progress came together in the invention of a new industry for the Chilean public, and in renewing the public's interest in the moribund affairs of the country's most vital export industry. Crucially, turning the industry inside out for the public (for the professional classes and the political elites) was only possible through the development of another industrial achievement in its own right: the setting-up of a formidable bureaucratic and administrative machinery, an intramural institution to which the industry obligatorily became accountable.

Until the late 1920s, production in the nitrate desert was organized around small nitrate refineries, known as *oficinas,* which hired individual labourers on a piece-rate basis. As early as 1878, there were already 165 refineries operating in the desert (Durruty, 1993: 30), creating a vast network of proto-capitalist outlets in the periphery of the world's driest wilderness. Distant as they were from both the coast and the capital city, Santiago, the social organization of the mining encampments came under the governance structures of the nitrate corporations. They were administered by the corporations as company towns (Stickell, 1979; Garces Feliu, 1999; Gonzalez Miranda, 2002), which assumed all responsibility on matters of institutional organization and social welfare, such as housing, health-care, schooling, safety, educational and recreational activities. The refineries' effective status as company towns meant that administrators had to do double duty as welfare managers and mining empresarios. From the 1920s onwards, the nitrate industry employed corporate welfare policies to redefine their relationship with the Chilean state (cf. Tone, 1997; Mandell, 2002), which effectively withdrew its presence from the desert and remained very much in the background until the 1960s. In this context, statist relationships took a corporate proprietorial form, with the corporations allocating and handing over housing to workers, taking charge of schooling and the routing of youth into vocational and professional education, 'owning' the courts and judicial system, drafting their own calendar of holidays and so forth. Hence the care with which the General Manager noted in the letter above that the industry should not be seen to disrespect the symbols of the nation-state. The magic of the state had somehow to be made not to disappear (Taussig, 1997).

Administrating the company-towns was a complex affair. Administrators had to embrace simultaneously an ethos of public service and a corporatist mentality. Thus the double standards and the aura of secrecy. The letter above was indeed stamped as 'reserved business' and we know from Weber that secrecy is one of the hallmarks of bureaucratic administration. 'Every bureaucracy', wrote Weber, 'seeks to increase the superiority of the professionally informed by keeping their knowledge and intentions secret. Bureaucratic administration always tends to be an administration of "secret sessions": in so far as it can, it hides its knowledge and action from criticism' (Weber, 1946: 233). It would certainly seem from the above that it is in the nature of bureaucracy to conceal its own openings, to distribute and make (public) knowledge travel so as to create its own communities of experts. A second set of examples will make this clearer.

Knowledge's Enclosures

The epistolary tradition gives away administration's secretive nature. Letter writing personalizes the institutional relation; it also turns into paperwork what could have been said verbally. It humanizes and formalizes all at once. It carries the source of its own invisibility in its visible format, alerting us to the 'public secrecy' (Taussig, 1999) that appears to characterize industrial relations. The letter format evinces the extent to which administration re-vertebrates institutional relationships through the elusive standard of its recursive protocol. Letters both extract themselves from, and induce, other letters, setting up a moving wall and benchmark for future correspondence – and future relations. Administration re-futurizes the institution.

On 12 April 1928, five school teachers from the Oficina Chacabuco wrote to the General Inspector of Oficinas at the headquarters in Oficina José Santos Ossa. Two regional newspapers had recently quoted the Provincial Director of Education on the appalling quality of teachers at Chacabuco. The teachers, on reading the piece, were dismayed. Their surprise was compounded by the fact that the Director of Education was only reporting what Enrique Peralta Torres, the Director of all Chacabuco schools, had previously told him. They felt offended and insulted – more so because Peralta Torres had publicly apologized to them in a previous communication. The teachers enclosed a copy of the letter, where the director explained to them that he had been mis-quoted when saying that staffing of educational centres in Chacabuco was 'insufficient', not 'incompetent'. The teachers, however, made it clear they did not believe the director's mistake was genuine. At a meeting with him on 9 April, the teachers had brought up the issue of his recent behaviour, to which Peralta Torres arrogantly replied that the teachers had not been behaving properly either: they had not been adhering to school regulations. The teachers brought the matter to the attention of the General Inspector because they knew the director's ways, and they felt it was important to state in writing their position before new developments start turning things around.

The teachers were right. In an undated letter sent to the Provincial Director of Education, presumably written shortly after the publication of the newspapers' pieces, Peralta Torres reported the shock that the articles brought to the families of Chacabuco. 'The people of Chacabuco', he says, 'were shocked ... yet honoured to have in me a Director with no ambition other than the preaching of truth and knowledge... They all acknowledge that my conduct was just and patriotic.'[5] In his missive the director mixed criticism of his peers with self-congratulatory passages and interspersed references to both men's friendship. He also bragged about the good spirit of companionship amongst teachers, despite their incompetence otherwise, in no small measure because of the frankness with which he dealt with and enlightened them.

Three further pieces of correspondence allow us to see the way the incident kept building up and at least one of the ways in which it was resolved. On 30 April, Peralta Torres sent a letter to the Administrator of Chacabuco. He alerted him to the fact that one of the schoolteachers, Carlos Riveros, did not go to work that day, a fault that made him eligible to being fired under Article 15 of the Law of School's Internal Regulations (Ley de Regimen Interno de las Escuelas). On that very same day, Peralta Torres prepared a second report on Riveros, which he also sent to the refinery's administrator. The report contained two pages of highly charged, personal vindictive statements. It opened, for instance, with the following assertion: 'I am writing to let you know that Mr Carlos Rivero, a teacher at one of our schools, is a pernicious character.' He further noted that Rivero is an 'enemy of the Educational Reform, for he always goes against the programme of progress, order and new educational tendencies that I seek to develop for the School.' The list of indictments went on: 'He does not comply with the law, nor the School's internal regulations'; 'he inflicts corporal punishment to the children'; 'he is poorly qualified, and his only ambition is to receive a salary for as little work he can do'; he was also accused of stealing tools that belong to the school and refusing to return them when found out. On 4 May the administrator of Chacabuco wrote to the General Inspector at José Santos Ossa endorsing the accusations levied against Rivero by Peralta Torres and requesting permission to suspend the teacher. On 16 May the administrator wrote again to the General Inspector asking for permission to appoint a new teacher following Rivero's resignation.

On 25 August 1931, the Chief Superintendent of Antofagasta's nitrate refineries, based in Chacabuco, wrote to the Antofagasta Region Police Commissioner. The letter was marked 'confidential' and referred to previous correspondence also stamped with that rubric.[6] The Superintendent wrote, he said, in anticipation of some possible developments. He had been following the recent activities of Luis Varas, director of the local public school. He was ignorant of, or preferred not to refer to, Mr Varas' actions inside the school but had good reason to believe they were like what had been made 'public'. He suspects such behaviour went 'against all principles of education, and against, also, the oficina's order and peace.' He was making the information available to the Commissioner in case the latter deems convenient to pass it on to the Provincial Director of Primary Education.

The Superintendent's letter was measured in the extreme, if not actually opaque in many of its passages. It was drafted in a bullet-point format and kept referring to external documents, some of which were enclosed. (The archives contained no copies of the enclosures.) He mentioned, for instance, a public speech that Varas gave on 30 July and to which he dragged his students, although it was otherwise poorly attended, and which was published in a local newspaper two days later. He enclosed a copy of the published version of the speech but provided no insight as to what it says. A second point refers to Varas sending telegraphs on behalf of a Workers' Committee (Comité Obrero) to a local newspaper reporting the death of a fellow worker. The dead man's widow was cited as saying she knew nothing about her husband being involved in political affairs. Copies of the telegraphs and a press clipping were also attached. Next, the Superintendent recounted Varas' self-appointment, on 21 August, to the Secretariat of the Frente Unico Civilista, Chacabuco's first ever political organization (which is disputable). The Superintendent was keen to stress the political character of the organization and to that effect enclosed copies of three flyers that were handed out to workers in the *oficina* over the past few days. Last, the Superintendent described Varas's refusal to involve his students in the celebration of the forthcoming fiestas patrias (national festivities), 17–19 September, on account of the risk of them being sunburned, when the matter came up at one of the meetings of the Committee for the preparation of the festivities. No documents were enclosed.

A letter's publics are sometimes not even public to the letter itself. The Superintendent wrote to the Commissioner so that the latter could, in turn, pass the report on to the Provincial Director of Primary Education. He referred his concern about the possible educational consequences of Varas's behaviour. He did not know much about pedagogy, but feared his behaviour inside the class might mirror what was 'publicly' known of him. Of course, what the Superintendent was really concerned about were the political activities of a potentially problematic labour activist and unionizer. The trouble was that politics made the magic of the state reappear; indeed, the police officers themselves were, or could be, re-agents in such potentially magical re-enchantment. And we know the company towns were corporately engineered to keep the state at bay. It was best, then, to fold the magic away, to hide it under, say, education. Best to 'enclose' the matter away, in copies of telegraphs, press clippings, flyers and other such items of 'public' knowledge and information. Public knowledge thrice enclosed: in education, in informational attachments, in a letter marked 'confidential'.

All too public was Peralta Torres' vicious attack on his staff, and on school teacher Carlos Rivero in particular. He told the press about it, and told the Provincial Director of Education, as well as the families of the children and, lest we thought he shied away from confrontation, told the teachers themselves at a meeting with them. The teachers retorted, of course. They wrote to the General Inspector of Oficinas, enclosing a self-inculpatory note written by Peralta Torres where he apologized for being mis-quoted. The enclosure was the teachers' claim to robust information. To little avail, unfortunately. The last we know is that the administrator of Chacabuco took the matter into his own hands and, 10 days later, Riveros resigned.

An analysis of the three cases of epistolary administration throws a number of issues into relief. Information and knowledge, for one, make their own claims to robustness through different expert and professional communities. The exchanges themselves conjure and legitimate the 'communities'. People appear and disappear in their institutional capacities at various junctures in the flow of information and every moment of institutional appearance delineates the field and definition of robust knowledge over which actors are staking out their expert claims. Robust or not, for instance, Chacabuco's administrator only took charge of the Peralta Torres' case when knowledge had reached a boiling point; he only made his appearance on stage when 'robustness', 'knowledge' and his own professional (expert) status intersected at a particular moment of 'public' recognition.

The boiling point of public recognition signals a second instance of contrast between the Varas and the Peralta Torres affairs. It concerns the other side – the hidden or secret side of that moment of public overvisibility: the moment when the question of the public itself became a secretive affair. Take the case of the enclosures. 'Not enough enclosures' is perhaps the thought that crossed the mind of Chacabuco's administrator when he looked at the teachers' correspondence and decided to endorse Peralta Torres' vindictive campaign instead. The teachers resort to 'external' evidence made for a feeble competitor against Peralta Torres' 'public' exhibitionism. Perhaps the teachers should have learned from Chacabuco's Superintendent and should have 'enclosed' their own enclosures. They should have turned the public robustness they were seeking inside out, and lightly touched their supply of information with an aura of secrecy. They should have reinvented and disguised themselves as an expert community by resorting to secrecy (cf. Cohen, 1971), exposing knowledge's self-enclosure in expertise and secrecy – and thus defending themselves.

This brings me to a third point, which is about administration's own sense of robustness. I said above that a letter's publics are sometimes not even public to the letter itself. No fetishism (of the letter) was intended. Gilbert Herdt criticizes Simmel for what he calls his 'miserly theory of secrecy' (Herdt, 1990). In this view, secrecy 'is about society, not about itself: secrecy must always refer to something other than itself, because, for Simmel, social reality is vested in society, not in the secret' (Herdt, 1990: 365). For this reason, Herdt prefers to view 'secret ideas and practices' as 'signs of themselves, not of what is outside of the secret system (i.e., "society")' (Herdt, 1990: 368). The efficacy of secrecy lies in its capacity to create its own natural kinds, in opening an ontological space for itself, that is, for the selves that inhabit it. Secrecy operates by fabricating 'an ordering of reality, a classification of persons, things and events, which must be kept separate' (Herdt, 1990: 368). It is (a new) society that lies dormant in the secret. It is secrecy that informs society, not society that animates the secret.

Secrecy's agency helps explain the letter's ignorance of its own publics. We have seen above how administration creates its own space for secrecy, a new secret order where certain categories of expert judgement, information and sociality take frontstage. Trade unionists, widows, children with sunburn, telegraphs, copies of political speeches and flyers populate the periphery of the Luis Varas affair. They extend the administrative agency of the letter into uncertain constituencies, into vague and porous publics. They open up a secret public for Luis Varas' political activities – conjured through surreptitious administrative attachments. Hence the fetishism of the letter and hence the power of the enclosure, the seduction of the attachment. The enclosure makes information robust by virtue of its very self-eclipsing, that is, by dis-placing its own claims to knowledge outside itself. It plays on the public/secrecy reversal to its own institutional advantage. Perhaps, in fact, that is the strength of administration: that it is a self-eclipsing mode of making knowledge available.[7] Administration allows for the nesting, indeed the 'interleaving', of knowledge (Deleuze and Guattari, 1994/[1991]: 50). It is a system of knowledge that, by definition, can only work when the knowledge itself is layered – 'admini-strated', stratified into and outside itself.[8] The exchanges of letters work as the structures through which institutional relationships re-vertebrate their conceptions of public knowledge. After all, administration is administration's most robust witness.

Conclusion

Marilyn Strathern has recently contrasted research and management models of knowledge and has observed that whereas the former takes knowledge's openings and bifurcations as its focus of interest, the latter is concerned with knowledge's closures and products (Strathern, 2004a). Whereas criticism is inherent to the former, it is derivative if not an obstacle to the latter. Making knowledge flow in the research mode requires very different 'organizational' skills from those deployed in the management mode. Knowledge management and the management of knowledge are not quite analogous operations.

Management, however, is central to both. Both researchers and managers need of administrative systems to put their respective visions on knowledge to work. For knowledge may not know where it is going but it needs to know where it is not going. Co-ordinating such dis-placements and dis-locations is what administration does best. Anthropology, for instance, goes nowhere without ethnography, which is our own, very special administrative toolbox: made up of fieldnotes, diaries, transcriptions of interviews, kinship diagrams, photographs, and so forth. (Not to mention the larger administrative system wherein anthropology itself moves, including the editorial process, peer reviews, book reviews, conferences, workshops and footnotes.) All anthropological analyses work as administrative systems of ethnographic knowledge. The analogy may come as an unwelcomed surprise to many academics, who are professionally sceptical of administration and dread it for the way it encroaches on their teaching and, often more importantly, on their research. No apologies are on offer, I am afraid. Administration, we have seen throughout the chapter, creates its own communities of experts. Anthropologists are no exception and they, too, may be dubbed administrative experts of a kind. We, too, preciously hold onto our administrative knowledge and have in recent times resisted the co-optation of ethnography by, for instance, cultural and media studies and sociology. No administration, no expertise.

Expert communities deploy their own tactics for 'public-ization', and these often involve redefinitions of the funds of knowledge itself. Anthropologists do this, Lakoff's pharmaceutical firms and psychiatrists do this, Hayden's ethnobotanists do it. When we look at institutions, however, expertise, knowledge and publicness have other ways of coming together. Institutions have ways about their publicness that are not strict reversals of 'private' interests: 'secrecy' might in fact do a better job here as analytical counterpoint to the notion of institutional publicness. A series of epistolary exchanges embedded in the history of corporate welfarism and state-business relations in Chile allowed us to see an example of this: how institutional public knowledge flows by, opening and closing gaps where secrecy redefines the legitimate spaces for institutional action. And administration played the trick. It was the system of administration that carried forward and embedded the institutional practice and redefinition of knowledge, professional expertise and publicness. Administration set the benchmark for the re-institutionalization of knowledge.

No doubt administration does not do this alone. Knowledge is contested and redefined in many other ways and domains. It would be absurd to claim a prerogative for administrative sociality. Indeed, it would be absurd to reify administration as a separate social domain altogether. Social relationships flow in and out of a variety of registers, of which the administrative is only one. In many respects, administration is just an emerging form of other social processes and assemblages (cf. Rabinow, 1999). However, though administration may not be a privileged, or indeed the dominant form through which knowledge is fashioned and redistributed in society, its legacy and imprint in industry today can hardly be disputed, whereas the force of industry itself in society is very much undisputable. Administration is to industry what industry is to society. Crucially, it is the institutional work that administration does that allows for the vertebral formation of knowledge across different social scales. There is more to industrial-cum-scientific knowledge than administration but, sooner or later, all scientific knowledge takes an administrative form.

Administration, then, does double duty as witness for and evidence in knowledge. Like Charon, the mythical boatman who ferried the souls of the dead to the underworld, administration's crucial role in the institutional trafficking of knowledge lies precisely in that it takes knowledge out – it carries knowledge over. It is the putting-out of knowledge that administration does best, indeed, that administration is all about. It is administration's own moment of 'publicization'. If we think hard about how industry and its institutions take themselves out into the public realm, I think we will find that administration comes up high in our list of achievers. If secrecy works as the analytical counterpoint to institutional publicness, we may find that administration does the reversible job for institutional public knowledge. Secrecy is to publicness what administration is to public knowledge.

The point on which I wish to conclude is that we – and I am thinking mostly of academics here – need to recognize the purchase of administration in the carrying forth of knowledge and do so for a number of reasons. First, administration is not going to go. We might therefore just as well make it go 'public'; that is, we might do well by starting to recognize its knowledge credentials and take stock of its productive hold over institutional life. Second, administration is a public institution:

a system of knowledge capable of eclipsing itself and thus of opening up and closing new and old constituencies. We all hate to read the small print but it is not unusual for new knowledge formations to emerge out of the over-standardized. Better to make sure that these work to our advantage by acknowledging their academic productivity. Third, recognition of administration's knowledge purchase should make us realize the futility of adding layers of administration to existing practices. More is not better. What we have already is (more than) enough – if only we knew how to put it to good use. Fourth, and of crucial importance for all research and knowledge institutions, administration should never be left in the hands of 'professional administrators' alone (Graham, 2005). All knowledge, I have argued throughout, is always and everywhere already administrated. A 'professional administrator' is therefore either someone who knows nothing except administration – which, of course, begs the question: administration of what? – or is already a disciplinary expert, an anthropologist, historian, engineer, and so forth, in which case administration needs only to be made productive. (A different question is whether in order to be made productive it needs to be made explicit too – and, if so, the kind of expertise that productive explicitness might need to mobilize.) The adage, in sum, is a simple but important one: administration is knowledge, even if knowledge is not simply administration.

Acknowledgements

Thanks to Rane Willerslev for his comments on an earlier version of this paper. I am very grateful also to the editors for their close reading of the paper and their feedback. The archival research on which this paper is based was partially funded by the British Academy, to which I am thankful.

Notes

1. I use the term 'administration' to denote the trafficking and systematization of knowledge in institutional contexts. Administration is different from bureaucracy and different from management yet permeates both, a point I elaborate later on in the argument.
2. 'Public' has recently emerged as new keyword of political rhetoric. My usage of the term follows Alastair Hannay's analytical unpacking: publicness as a conceptual fund wherein notions of liberal individuality, a space of intellectual and market exchanges, and new social and political audiences take form (Hannay, 2005).

3. On the distribution of self-similar properties at different scales, see Strathern (1991). For an ethnographic example of administration's (in this case, accountancy) fractal qualities, see Maurer (2002).
4. Archivo Nacional de Chile, *Catalogo Inventation del Salitre, Lautaro Nitrate Company,* Vol. 48, pp. 33–4.
5. All correspondence, unless otherwise noted, from Universidad Catolica del Norte, *Archivo Historico, Oficina Chacabuco,* 1928, various unclassified rolls.
6. Universidad Catolica del Norte, *Archivo Historico, Sindicatos,* unclassified roll.
7. F. G. Bailey, in his classic monograph on the folklore of academic politics, makes a similar observation about secrecy and its place in the 'backstage' of the politics of administration: 'secrecy and duplicity ... [is] a rather sophisticated strategy which is designed to allow the administrator simultaneously to run with the hare and hunt with the hounds' (Bailey, 1977: 54).
8. Hierarchy and layering are indeed part of the etymology of 'administration': from *ad-ministrare*, the action of ad-ministering or serving in office, serving as a minister: servant, subordinate, formed after the correlative *magister,* master (Oxford English Dictionariy, www.oed.com).

Bibliography

Bailey, F. G. (1977), *Morality and Expediency: the Folklore of Academic Politics,* Oxford: Basil Blackwell.

Becher, T. Trowler, P. R. (2001), *Academic Tribes and Territories: Intellectual Enquiry and the Culture of Disciplines,* Buckingham and Philadelphia: The Society for Research into Higher Education & Open University Press.

Brenneis, D. (1994), 'Discourse and Discipline at the National Research Council: a Bureaucratic Bildungsroman', *Cultural Anthropology,* 9: 23–36.

Cohen, A. (1971), 'The Politics of Ritual Secrecy', *Man,* 6: 427–448.

Deleuze, G. and Guattari, F. (1994 [1991]), *What is Philosophy?* Trans. H. Tomlinson and G. Burchell, New York: Columbia University Press.

Douglas, M. (1986), *How Institutions Think,* Syracuse, New York: Syracuse University Press.

Durruty, A. V. (1993), *Salitre, harina de luna llena,* Antofagasta: Norprint.

Etzkowitz, H., Webster, A. and Healey, P. (eds) (1998), *Capitalizing Knowledge: New Intersections of Industry and Academia,* Albany, New York: State University of New York Press.

Garces Feliu, E. (1999), *Las ciudades del salitre,* Santiago de Chile: Origenes.

Gellner, D. N. and Hirsch, E. (eds) (2001), *Inside Organizations: Anthropologists at Work,* Oxford and New York: Berg.

Gonzalez Miranda, S. (2002), *Hombres y mujeres de la pampa: Tarapaca en el ciclo de expansion del salitre,* Santiago de Chile: LOM Ediciones.

Graham, G. (2005), *The Institution of Intellectual Values: Realism and Idealism in Higher Education,* Exeter: Imprint Academic.

Handelman, D. (1981), 'Introduction: the Idea of Bureaucratic Organization', *Social Analysis,* 9: 5–23.

Hannay, A. (2005), *On the Public,* London and New York: Routledge.

Hayden, C. (2003), *When Nature Goes Public: the Making and Unmaking of Bioprospecting in Mexico,* Princeton and Oxford: Princeton University Press.

—— (2004), 'Prospecting's Publics', in K. Verdery and C. Humphrey (eds), *Property in Question: Value Transformation in the Global Economy,* Oxford and New York: Berg.

Herdt, G. (1990), 'Secret Societies and Secret Collectives', *Oceania,* 60: 360–81.

Herzfeld, M. (1992), *The Social Production of Indifference: Exploring the Symbolic Roots of Western Bureaucracy,* Chicago and London: University of Chicago Press.

Heyman, J. M. (1995), 'Putting Power in the Anthropology of Bureaucracy: the Immigration and Naturalization Service at the Mexico-United States border', *Current Anthropology,* 36: 261–87.

Lakoff, A. (2005), 'The private life of numbers: pharmaceutical marketing in post-welfare Argentina', in A. Ong and S. J. Collier (eds), Global Assemblages: Technology, Politics, and Ethics as Anthropological Problems, Malden, MA, Oxford, and Carlton: Blackwell Publishing.

Lundin, R. A. and Söderholm, A. (1998), 'Conceptualizing a Projectified Society: Discussion of an Eco-institutional Approach to a Theory on Temporary Organizations', in R. A. Lundin and C. Midler (eds), *Projects as Arenas for Renewal and Learning Processes,* Dordrecht: Kluwer Academic Publishers.

Mandell, N. (2002), *The Corporation as Family: the Gendering of Corporate Welfare, 1890–1930,* Chapel Hill and London: University of North Carolina Press.

Maurer, B. (2002), 'Anthropological and Accounting Knowledge in Islamic Banking and Finance: Rethinking Critical Accounts', *Journal of the Royal Anthropological Institute,* 8: 645–667.

Miller, D. (2003), 'The virtual moment', *Journal of the Royal Anthropological Institute,* 9: 57–75.

Miller, P. and O'Leary, T. (1994), 'The Factory as Laboratory', in M. Power (ed.), *Accounting and Science: Natural Inquiry and Commercial Reason,* Cambridge: Cambridge University Press.

Nowotny, H., Scott, P. and Gibbons, M. (2001), *Rethinking Science: Knowledge and the Public in an Age of Uncertainty,* Cambridge: Polity Press.

O'Neill, O. (2002a), *Autonomy and Trust in Bioethics. The Gifford Lectures, University of Edinburgh 2001,* Cambridge: Cambridge University Press.

—— (2002b), *A Question of Trust. The BBC Reith Lectures 2002,* Cambridge: Cambridge University Press.

Ong, A. and Collier, S. J. (eds) (2005), *Global Assemblages: Technology, Politics, and Ethics as Anthropological Problems,* Malden, MA, Oxford and Carlton: Blackwell Publishing.

Power, M. (1994), *The Audit Explosion,* London: Demos.

Rabinow, P. (1999), *French DNA: Trouble in Purgatory,* Chicago and London: University of Chicago Press.

Schwartzman, H. B. (1987), 'The Significance of Meetings in an American Mental Health Centre', *American Ethnologist,* 14: 271–94.

Shore, C. and Wright, S. (1999), 'Audit Culture and Anthropology: Neo-liberalism in British Higher Education', *Journal of the Royal Anthropological Institute,* 5: 557–75.

Slaughter, S. and G. Rhoades. (2004), *Academic Capitalism and the New Economy: Markets, State, and Higher Education,* Baltimore and London: John Hopkins University Press.

Stickell, A. L. (1979), *Migration and Mining: Labor in Northern Chile in the Nitrate Era, 1880–1930,* Unpublished PhD thesis: Indiana University.

Strathern, M. (1991), *Partial Connections,* Savage, MD: Rowman & Littlefield.

—— (ed.) (2000), *Audit Cultures: Anthropological Studies in Accountability, Ethics and the Academy,* London: Routledge.

—— (2004a), *A Community of Critics? Thoughts on New Knowledge.* Huxley Memorial Lecture, Wednesday 8 December, University College London.

—— (2004b), 'Working Paper Two. Commons and borderlands', in M. Strathern, *Commons and Borderlands: Working Papers on Interdisciplinarity, Accountability and the Flow of Knowledge,* Wantage: Sean Kingston Publishing.

—— (2005), 'Robust Knowledge and Fragile Futures', in A. Ong and S. J. Collier (eds), *Global Assemblages: Technology, Politics, and Ethics as Anthropological Problems,* Malden, MA, Oxford and Carlton: Blackwell Publishing.

Taussig, M. T. (1997), *The Magic of the State,* New York and London: Routledge.

—— (1999), *Defacement: Public Secrecy and the Labour of the Negative,* Stanford: Stanford University Press.

Tone, A. (1997), *The Business of Benevolence: Industrial Paternalism in Progressive America,* Ithaca and London: Cornell University Press.

Weber, M. (1946), 'Bureaucracy', in H. H. Gerth and C. W. Mills (eds), *From Max Weber: Essays in Sociology,* New York: Oxford University Press.

Whyte, W. F. (ed.) (1971 [1946]), *Industry and Society,* Westport, CT: Greenwood Press, Publishers.

–3–

Rationality and Contingency

Rhetoric, Practice and Legitimation in Almaty, Kazakhstan[1]

Catherine Alexander

Introduction

During a conversation in 2001, Murat, a government economist in Kazakhstan, stood to explain why the Soviet system had failed and market-based economics appeared to be so successful. He picked up a large stapler and held it out at arm's length. 'In the Soviet times we thought that only Marxist economics were possible; this was how everything had to be. But then with perestroika and afterwards we realized we had not known the truth. This is what I tell my students now.' He dropped the stapler with a crash onto his desk, 'I tell them that the market is inevitable like the law of gravity.'

James Scott (1998) famously opposed two modes of knowledge: abstract, universalizing rule-bound techne and the practical knowledge of metis that is drawn from an understanding of local specificity. These modes were further identified with the respective analytical categories of 'the state' and 'society', both of which are themselves abstractions. It is a powerful model, particularly when applied to the high modernism typically associated with state socialist[2] experiments in the twentieth century; but a model is also just that: something (see Harbison, 1991) that is at once paradigmatic, in the sense of an ideal type, and suggestive of something greater than itself. What I explore here is, in part, that very suggestiveness: the rhetorical and instrumental freight carried in different circumstances by the idea of universalizing rules based on right reason versus the embedded wisdom of pragmaticism. This chapter then aims at uncovering something other than the problems facing the process of translation from idea to practice, from laboratory protocols or bureaucratic schema to the world outside, with which the literature on science and technology studies concerns itself. Moreover, as this chapter tracks ideas of scientific rationality from the Soviet era to the contemporary period, the legitimating function of revolutionary Marxist logic, grounded in historical determinism, as retrospective explanation and justification is shown to continue in the twenty-first century. According to this logic, as Murat's speech above shows, the Soviet system fell and a market-based economy emerged as an inescapable result of natural law.

The context through which I have chosen to explore first, how far Scott's thesis holds and, secondly, the rhetorical work to which rationality was, and is put, is that of a local-level state administration: the municipality of Almaty,[3] Kazakhstan's former capital. I follow this from Soviet to contemporary times via the planning function (as illustrative of techne in the world) and the officials associated with it. The periods of Soviet Almaty referred to here, as memories and accounts weave backwards and forwards through time, are roughly as follows. Alma-Ata, a small town in the south east, was made the capital of Soviet Kazakhstan in 1929. Over subsequent decades, the city grew rapidly in size for the reasons described below, but it was not until the 1960s that large-scale construction began, both of housing, and public buildings, which continued to change the city's appearance rapidly until money petered out in the late 1980s. These were variously known as the 'stagnation years' under Brezhnev, or Alma-Ata's 'golden years' under its First Secretary, Kunaev. In 1991, the Soviet Union collapsed and a mass-privatization programme began shortly afterwards. My main body of data is taken from fieldwork 2000–5, although I have also drawn on historical accounts, archive photographs and newspapers, and semi-fictional accounts of 1930s Alma-Ata. Throughout this entire period, both Soviet and after, conflicting narratives may be detected: those celebrating Soviet modernist progress and those rejoicing in and then mourning the natural abundance for which the pre-war Alma-Ata had been famed and was later severely reduced. I should emphasize that the same informant, whether official or citizen could quite happily occupy both positions in one conversation.

The reasons for choosing the municipal administration for my case study are twofold. First, in the persons of city bureaucrats are combined the contradictions of the modern world; they are at once the faceless implementers of legal-rational codes and plans[4] – and citizens of the urban world on which they act and in which they live (see also Hart, 2005). On leaving the relative calm of offices, statistics and projections they must now negotiate, for example, the sight and sound of Tajik refugees begging in the streets, unregistered cars belching exhausts from adulterated fuel, and the tall building developments shooting up around the city, interrupting the once famous skyline of trees. The disjunction between plans and reality was ever thus: small, self-built dwellings crammed into spaces between carefully planned Soviet city districts are a constant reminder that multiple ideas and practices of the city always co-existed. While engaged in planning future city developments therefore, officials, especially during the Soviet period, drew both on ideologically inspired schemas and an intimate familiarity with the local environment.

The second reason for choosing the local urban planning function is to examine what actually happens to these epistemological modes at the level of implementation from two angles: first how the rhetorical work to which 'rational science' was put during the Soviet era changed over time, but is still used as a logical framework to explain the advent of a market-led economy. The second angle is the friction between modes of knowledge and practice within different levels and interest groups within (and beyond) the administration both during the Soviet period and after. Just as the socialist economy was famously supported by informal structures and 'shadow' or 'black' economies (Sik, 1992, 1994; Verdery, 1996; Kurkchiyan, 2000), so might one see the driving force of revolutionary logic, apparently irrefutably based on physical laws, as being shored up by local knowledges. Latterly, as in Murat's case, the reverse has happened: stark social and economic changes ushered in with independence, have been retrospectively 'rationalized' in some quarters by that same logic.

The pragmaticism of local knowledge, however, is worth dissecting a little further because it appears in several forms, circumstances and relations to standardizing bureaucratic knowledge. In the first instance, officials' experience of the immediate environment might contradict the sagacity of centrally-issued plans premised on uniformity (French, 1996, see also Ferguson, 1994), and after the 1940s, the dictates of powerful industrial ministries in Moscow (Beisonova, 1998). Thus, in 1930s Soviet Alma-Ata, low buildings were angled to the streets to maximize windflow through the city in recognition of the sluggish circulation caused by the depression in which the city was built. As local planning power drifted towards Moscow and Leningrad in the 1940s, however, the reverse happened and pattern housing blocks were built of materials common across the Soviet Union, but woefully inappropriate in extreme continental climates such as Kazakhstan.[5] The second instance of metis mingled with techne, is more imbrication than opposition; there are at least two cases where this might occur. One might think of the strategies employed by workers and factory directors in a soft budget regime (such as padding stock inventories) to ensure that production targets could be met even when the supply part of the overarching plan was deficient (Verdery, 1996). Here, rules failed when translated from schema to the world because of unaccounted-for variations or deviations from assumed norms, but it was local practical knowledge of how *the system* itself worked in fact, as well as of those local variations, that enabled plans to be patched up and hobble along. Another version of this instance of metis sustaining techne is the example described below of larger scale contingent events over-riding plans based on statistical projections. World War II, for example, shattered demographic and other forecasts. In common with other Soviet cities (French, 1996), Alma-Ata's population consistently outstripped 'scientific' estimates in the latter half of the twentieth century, providing urban administrators with a continual crisis of providing basic accommodation, healthcare and education. Quite different criteria of need were used to determine the priority of such building developments from the plans and building regulations pitched at creating the best of all possible worlds on a rational basis. Local urban planners had one more form of contingency to cope with when trying to devise and implement grand metropolitan designs; this was the unforeseen event created directly by the kind of large-scale project Scott (1998) describes, which was monocular in focus and failed to consider potential knock-on effects. One example for Alma-Ata here is the constant demand for housing provision for released prisoners and their families from Kazakhstan's gulag network.

The last point to explore briefly in this introduction is the plasticity of the ideas of 'science' and 'rationality', which had such a powerful shaping force on Soviet, indeed post-Soviet, ways of thinking and acting. Marx saw in technology the means of harnessing nature in the service of mankind, much as technologies of culture could overcome the more obdurate elements of individualist human nature. To understand and thereby use science was to uncover positive laws, which transcended localized moralities of right and wrong with universal inevitability. But, there are already a number of things going on here that need to be unpacked a little. The *goal* of science in the attenuated form it began to acquire from the seventeenth century onwards, is the understanding of physical laws that drive the world. The *method* to gain such comprehension is the induction of universal rules from the observation of particular phenomena, usually relatively isolated and usually in a controlled or simulated environment. Scientism, which underpinned the high modernist project of socialism in the twentieth century, if not in so many words, made two huge conceptual leaps from this goal and method in terms of extrapolation and applicability. First, there was the belief that principles of behaviour and structure found at one scale, in one domain, in one timeframe and under a particular set of circumstances could be actually, not just metaphorically, extrapolated to others. Thus, for example, organic transformations over millennia segued into the possibility of cultural, social and political change in relatively short order. The interventionist notion that laws could also be applied across domains to hurry up an otherwise inevitable process was the second scandal of this faith. Soviet planning had a curious relationship to these ideas of historical inevitability and science. On the one hand, rational thought dictated the best possible way to live; on the other the determinism of some plans was rejected under Stalin as attempting to counter certainty.

The form that scientism took in the Soviet Union was also tinged with a chiastic flavour that clearly pointed to a full stop. That endpoint – the realization of communism – was portrayed as a condition of harmony within and between social, political and natural domains: a place for everything and everything in its place. There was thus an awkward combination of ideas about rational order, which did not logically fit together. Scientific rationality as a method in this framing was the tool to unravel observable cosmological laws that were themselves driven by the prime mover: a logical law of the inherent Manichean nature of things. This intrinsic duality of matter and thought effected (ironically) a boom bust blind progression onwards. Quite how the dialectical method with its implication of ceaseless change slotted in with a future finitude and cessation of movement was not crystal clear. Nevertheless, as the rest of this chapter shows, though ideas of a right order played an important role in Soviet self-exegeses from top to bottom, there were quite fundamental differences in the nature of that coherent, rational order that was summoned forth by citizens and officials respectively.

The rest of this chapter is divided into two parts. The following section takes Almaty as a case study through which to examine these different forms of rational order. The architectural compositions of the city centre, and one area in particular, are considered through the official ideas of rational and social harmony that they suggest, together with the ways in which contemporary Almaty's citizens speak of different orders in and through the same places. The second part considers the scientific method as a means of conducting bureaucratic affairs, specifically, planning the city and directing it towards a goal of civic harmony. The problems for Almaty's city administration after 1991 changed; nevertheless there was still a clash between rational orderly planning based on empirical observation and projection on the one hand, and the constant overflow of events from the wider world, whether the higher governmental bureaucracy or commercial interests that trampled over the careful plans.

The conclusion considers these various forms of rationality and contingency in the light of the succession of Soviet Constitutions, the latest of which announced that a condition of harmonious dwelling in the world had in fact been achieved. As the following suggests, the impression for many of my informants was in fact quite the reverse: that the instrumental scientific method had served to destroy an existing harmony.

Part I

Almaty is tucked in the south east corner of the country where a spur of the Tien Shan mountain range marks the border between Kyrgyzstan and Kazakhstan. In tune with the 'Kazakhification' programme directed towards the state bureaucracy after 1991, the municipal administration was renamed the 'Akimat', which is also the name, by extension, of the administration's main offices. The principal building, previously home to the Parliament of the Soviet Socialist Kazakh Republic (SSKR) since the 1980s, is a huge, white edifice that rears up behind a vast, open space. This is where, in the mid-1980s, one of the main axial roads running east to west, was widened into a place for marching and public celebrations of key dates in the official Soviet calendar (Wanner, 1998). In the space of only 25 years, the name of this place slid from New Square to Brezhnev Square and then, in the aftermath of the Soviet Union's fall and Kazakhstan's inauguration as an independent republic in 1991, Republic Square.

Photographs from the 1980s show surrounding buildings draped with gigantic banners of Lenin and Marx; whirling, beaming dancers tap-dancing out a red star; and exhortatory placards urging workers to unite, beneath which there are small tableaux of workers in nineteenth-century costumes. The square is overflowing with activity. Curiously, the relentless gaiety and staginess of the smiles turned to the photographer seems to lock the pictures into a deeper past than the scribbled dates on the back indicate.

Now, just as then, pedestrians who stand at the lower, northerly edge of the square have their sight drawn up towards the building where, behind the domineering, slab block façade, there is a breathtaking view of snow-capped mountains, a backdrop to the highest administrative construction in the city, in all senses. The theatricality is precise and deliberate. The southern half of the city slopes up towards the mountains. With all the artful, apparent carelessness of an eighteenth-century garden, vistas were minutely planned in the 1970s and 1980s by local architects and town planners so that the pedestrian toiling up the tree-lined boulevards of the city's central grid of roads would pause and look up to see glimpses of splendid, civic constructions at the upper end of each stretch of road: the Opera House, the Academy of Sciences, the National Library. Most of these structures were rendered even more imposing, first by an open expanse of grass or small square to the fore that framed and accentuated the centrepiece and then by the geographical orientation so that the building loomed out of its shadow, edged in bright sunlight.

Dinmukhamed Kunaev, First Secretary of the SSKR from 1960 to his deposition in 1986, was a civil engineer by training and much beloved by the architects both of his day and the present. The reason is simple. Under his jurisdiction, Alma-Ata grew from a pretty but small, backwater town into a city. His friendship with Brezhnev, dating back to the latter's stint in Kazakhstan on the Virgin Lands Campaign[6] in the 1950s, encouraged the flow of money from Moscow directed towards magnificent projects quite explicitly designed to transform Alma-Ata into a metropolis worthy of being the capital of the SSKR. As middle-aged architects now happily and rather wistfully recall, Kunaev took a direct interest in promoting the design and execution of daringly new projects that tried to combine splendour with a sense of local architectural features. This was no easy brief where traditional society was nomadic and therefore placed less stress on the built environment. Successful architects were also treated well materially by Kunaev, which possibly accounts for some of the nostalgia. But it was in the development of the then upper reaches of the city above the key axial road, Abay Prospect, that Kunaev's full architectural ambitions were unleashed. Despite the initial support from Moscow, however, for a grand set piece of a stepped avenue, bordered by fountains, opening out into an immense square for Soviet celebrations and so on up to the great white Parliament building, the Moscow planners deemed the draft plans too ambitious in scope and size: they even dwarfed Red Square. Clearly, this would not do. The Chief Architect of Soviet Alma-Ata in the 1980s, who was telling this story in 2004, recounted how Kunaev was told in no uncertain tones to reduce the scale of the plans. After all, manifesting the centrality and supremacy of the Soviet administration in Alma-Ata was one thing, but it was quite another when a far-flung satellite Republic threatened to overshadow the heart of the Soviet empire. That, quite literally, would be out of order. Accordingly, the plans were trimmed back and the grand (but not too grand) architectural spectacular was created at the then highest point of the city.

Thus, the Akimat tops off the display of architectural pyrotechnics of the decades when the political leadership sought to establish Alma-Ata as a republican capital, not just in name but in visual splendour, elaborating on a close scientific knowledge of local geography, geology and climate with the all the perspectival tricks of the town planners' trade. Despite its unpromising location, tucked into a dip at the base of the Tien Shan and sitting on sandy soil riven with underground streams, Alma-Ata was made magnificent. A phenomenological sense of harmonious order was wrought upon the individual citizen who took his or her place in the rational whole. State planning, through technological methods and abstract knowledge, was married to an intimate knowledge of the local environment to maximize the effect and experience of harmony within and between the social, built and natural orders. Enthusiastic contemporary accounts, as in this quotation cited by local historian, Ilya Malyar (1976), indeed bear witness to the planners' achievement in weaving together nature and human needs, through technology, to produce a city that delighted the senses and answered ideological requirements.

> Huge parks, wide streets, houses drowning in greenery: all of this turns Alma-Ata into a garden city which was built in accordance with social political directions for the creation of healthy conditions for the bright and happy life of human beings.

At least, that was the official version.

Walking through this area now with citizens, whether bureaucrats, doctors, market traders or architects,[7] prompted slightly different accounts, some recycled from parents' and grandparents' family stories, some recalling childhood memories. Nearly all of them spoke of the wonderful collective farm Gorny Gigant (Mountain Giant) that had reached south from where Abay Prospect is now, away up to the mountains. Then there were apple orchards with mile upon mile of beautiful blossom that produced the amazing local hybrid apple, Aport, which was there to be freely plucked from the trees by citizens. Little dachas were set amongst the fields and orchards, where now there is a large, concrete headquarters for a television company and traffic screams through Republic Square. In the 1950s, people tell you, there was only the sound of birds and gentle, wild animals that would come right up to the dachas' front porches. More than this, there was the rich, thick, black soil of the collective farms ringing the city. Soil that, as a retired sewerage engineer observed bitterly, was ploughed up to build the masses of concrete apartment blocks in the 1960s. 'Soil that took the earth hundreds of thousands of years to make centimetre by centimetre,' he said, 'destroyed in a few years by those shoddy, crumbling blocks. A terrible crime.'

In other words, citizens persist in seeing past worlds beneath the concrete pavements their feet are walking over. This particular past is strikingly Paradisical, timeless and whole, ripped apart by statist diktat: by directive, consumptive manipulation of nature. As a coda, it is worth pointing out the discomfort of someone who was standing by, listening to the engineer's jeremiad. Her mother had previously told me how, as a three year old moving to one of these new two-room apartments in the 1960s with her parents, she had run from room to room shouting and jumping with excitement at the space it afforded; the last place she had lived in had been a single room for six people with a communal kitchen and bathroom. Incompatible Edens have always existed.

Revisiting this highly orchestrated area from a number of angles suggests a range of alternative understandings of the social harmony manifested in this grand architectural composition. The first counterpoint is through those memories alluded to above, which all but erase the present environment and future possibilities in recollections of past worlds and different harmonies. Quite apart from such stories, though, Republic Square is oddly desolate other than for the traffic accelerating through the middle stretch. This is not just because it is vast. In 1986, there was a riot against the central imposition of the Russian First Secretary Gennadii Kolbin in place of the much loved Kazakh, Kunaev. No-one knows precisely how the riot began or ended, but there were deaths and they haunt the place. One old Kazakh woman living with her son's family on the outskirts whispered conspiratorially to me that thousands of Kazakhs had been murdered and tipped in a pit dug far away in the steppe. This is unlikely but it fitted with a particular genre that cast the state as a unified 'they' from Soviet to current days, from national to local government: unpredictable, savage, disturbing the even tenor of local balance.

Citizens approaching the Akimat across Republic Square in 2000 said they felt intimidated by its size, even though they were shunted away from the grand frontage round to a small side entrance. The splendidly carved double doors of the front elevation leading to a vast, marbled atrium only swung open for those whose appointments with officials had been set-up in advance or for visiting dignitaries. Most citizens who had grievances or, more typically, were gathering together the cumbersome portfolio of documents, stamps and signatures needed to carry out anything in the city, had to shuffle though the easterly door. There they joined a long chattering queue in a small room to use the single working internal telephone. This allowed them brief access to an official whom, they hoped, was both in and willing to instruct the security guards to allow the visitor to visit their department.

It is common, now, to hear people say that officials have become more remote and that decisions concerning the city are just announced in newspapers rather than being discussed, as happened before. Indeed, the 25-metre-wide Abay Prospect is punctuated by clusters of architectural compositions, all created via Union-wide open competitions in the 1970s and 1980s. Iron notice boards along main roads, which had served to carry posters publicizing such architectural competitions and plans for future city developments, now stand empty except for scraps of paper advertising apartments for sale or the occasional bill for forthcoming films or theatrical events. Whether or not decisions about Soviet Alma-Ata were ever actually affected by citizens' opinions is almost irrelevant, there was at least the publication of notices, competitions and letters as if they were. Many Soviet citizens may indeed have been cynical about the workers' state (Yurchak, 1997 following Sloterdijk, 1987) but its replacement version has also caused older residents and officials alike to say that good things have been thrown out with the bad. Citizens' understandings, always fragmentary, have now become incomplete in a different way: certain shared fictions between state and citizens, which served to veil dissonances in the official line, have been pulled away leaving only frayed disconnections.

Part II

To revisit the Akimat again, but this time to go behind that daunting façade, is to find another approach to orderliness and system in the practices and understandings of local officials. Just as city residents often homogenize state activity, whether talking about the Soviet or contemporary administrations, so distinctions are also rarely made between local and central government: it is all the state. But this view is misleading. To an extent, before independence, things *were* stitched together via Party affiliations, and officials *did* move between different echelons, ministries and departments. But, after 1991, the relatively seamless official organization was multiply fissured. The connection to Moscow was lost. In 1997, the political capital moved away from Almaty to Astana, taking with it all the ministers with whom bureaucrats had routinely conducted face-to-face negotiations, confident that they were all part of the same overarching system and sharing a common bureaucratic background. Now, such discussions are unsatisfactorily mediated by telephone or entail a long journey. Some senior officials impatiently speak of the impossibility of trying to do business with central government now that ministries are stuffed with what are pejoratively called the Harvard Boys rattling off the new truths from positivist economics textbooks. One Akimat official in his sixties gestured despairingly in the vague direction of Astana, 'How can I do business with people like that?' he said, 'Young girls with skirts up to their thighs.'

These are the other differences. The gulf between generations is often more than simply age and length of work experience. Young officials, freshly weaned from university and recently imported textbooks, lack the breadth of experience prized amongst senior officials, who now find their practical knowledge at less of a premium than hot-off-the-press *Wissenschaft*. These older officials were socialized into the Soviet bureaucracy, which promoted a particular way of doing things for a particular purpose. Although that has been formally lost, many informal practices continue as the only way of getting the job in hand done. But, for some, it is hard to become accustomed to the loss of familiar methods and goals.

The Akimat itself, in other words, now operates through a babble of competing local knowledges and world views; there is no longer a singular received method of how the affairs of the bureau should be conducted, nor what the role of the public sector is *vis-à-vis* the citizens of Almaty. One point that comes out of this is that although privatization and such buzzwords are the order of the day, it is quite another thing when it comes to changing the minutiae of daily business practice, perhaps because it is so intimate and touches at the core of the previous bureaucratic cosmology: a way of seeing and acting on and in the world, a way of understanding how the world ought to be.

As parts of the Akimat were snapped off and turned into private businesses, remaining officials said how hard it was to understand how they were supposed to carry out their jobs when it was suddenly apparently corrupt to conduct business with someone they had worked alongside for years and trusted, but who was now, through the wave of the privatization wand, somehow on the other side of the public/private chasm. In the circumstances, it was something of a comfort to continue collecting the statistics on which the new 2002 General Plan for the city was based.

Now, unless small in scale and in a controlled environment, few plans are ever fully implemented. But, what is remarkable, looking at previous city plans for areas beyond the immediate city centre, is the extent of the gulf between these blueprints and the colourful jumble of small private dwellings that form the outskirts, often with no heating and street standpipes for water supplies. More than that, as the Leningrad Planning Department's 1949 General Plan for the

City of Almaty shows, these still representations show an orderly city, a city of grand boulevards and, with more than a hint of classical regularity, a city that was open and under the careful management of a central, harmonizing authority. Indeed, local citizens usually said this must be a picture of Ancient Rome or Greece when I showed them an A4 photograph of a painting taken from this Plan. In effect, this city, this city plan, was a city to think with, as were all previous and subsequent plans, an ideal projection, rather than a city to live in with all the tumbling cacophonous complexity of urban life.

One simple answer to why these plans were never implemented in their entirety is that, in each decade, events beyond the control of the city management scrambled their precise demographic forecasts, presenting the impoverished and under-skilled administration with a desperate demand for housing, schools and hospitals, whether or not they conformed to building regulations, let alone the whole plan. In one typical but telling example, Alma-Ata's population in the mid 1960s already exceeded the previous decade's demographic projections for the mid 1980s.

Thus, mass collectivization and famine in the 1930s saw starving nomads pour into the city (Carrère d'Encausse, 1994: 260). In the 1940s, factories and their personnel were relocated from frontline territories at war to the peaceful Soviet hinterlands of Alma-Ata. In the 1950s and 1960s, tent cities sprouted on the steppe to accommodate workers flooding in for the Virgin Lands Campaign. Each decade, the city strained to accommodate a sharply escalating population until, in the 1960s under Khrushchev, hundreds of standardized, low-quality apartment blocks were flung up as a response to this unforeseen need – one directly created through the process of centralized planning that was to redeem mankind from the vagaries of chance, to paraphrase Marx. These were the apartment blocks, built in small clusters, to which the engineer referred when commenting on the destruction of the agricultural land on which they were built.

Clashing planning directives from different ministries and different scales were not the only problem: Soviet regulations described an ideal. They set far higher standards than comparable regulations in the West; many construction and environmental regulations still follow this tradition.[8] As with the broader plans, such regulations were based on close observations of physical phenomena and were geared towards perfection, the ultimate telos. Consequently, they were unachievable once translated to an environment of infinite variety and practical constraints. More perhaps than most citizens, officials worked, and still do, according to the double vision of ideal process on the way to still finitude, and the hasty pragmatic compromises forced upon them by contingent incidents.

Intriguingly, the Akimat still produces city plans based on statistics. Before 1991, the gap between goal and what actually happened can partly be explained by swamping contingent demands, generated either *by* the planning process itself or from beyond the environment it could control. Now the plan is divorced from both telos and genesis: nobody is sure where they are going and the basic data are dubious. The new city plan, I was told by its Chief Co-ordinator in 2000, was simply to follow the market. While theoretically sticking close to the Soviet urban planning mantra of separating residential zones, the rules of the market would determine the direction of the city's expansion. This has not been an entirely happy exercise to date, not least because as the plan went to press in 2000, it was discovered that land earmarked for future enlargement had already been built on by semi-legal developers. Rather than eject the entrepreneurs, revisions to the plan were issued – a practice that has continued. The plan, the careful, rational plan, was no less insured from external events than it had been before, even if they were produced by different circumstances.

Despite all this, officials and academics still laboured to produce the plan from its supporting layers of detailed environment, demographic and economic forecasts; these, however, were based on extremely doubtful statistics. Most current urban demographics in Almaty are unreliable simply because the previous system that pinned people down with official documentation has dissolved (see also Humphrey, 2002). After 1991, the city was freshly bulked with migrants from economic and environmental disaster areas, people who vanished into Almaty leaving no documentary trace behind and so, in effect, becoming bureaucratically invisible. Still, however, the statistical machine churned through its allotted tasks producing, for example, statistics on morbidity, despite the huge problem of ex-prisoners infected with TB refusing to register once they have been discharged from prison; increases in vehicles, although few of the elderly cars that throng the once-silent streets are registered; and migratory movements in and out of the city, although many have neither identity documents nor resident permits.

In conversation, officials demonstrated that they were well-aware of these discrepancies yet, reduced to bureaucratic myopia – only able to see objects through their documentary representation – they continued to use such statistics as the basis of their work. The rational process of delivering outputs from statistical manipulation continued even when a few minutes wandering along the city's streets would suggest something must be awry with the source data.

Thus there are multiple life worlds that not only co-exist within the Akimat, but are also negotiated on a daily basis by each official. Marx observed that, 'the bureaucrat has the world as a mere object of his action' (Marx, 1970: 48) but, at the same time, these officials also live in the world that they deal with through such orderly abstractions. The ideology of the market unbound, which officials had to espouse at work, resulted in, as the Head of Construction observed, load-bearing walls being ripped out in over-hasty conversions of ground-floor flats to shops. 'They're like termites, these businessmen,' he said as we strolled through the streets, 'destroying our city. The apartments above those shops are left to balance on a pillow of air.' It is worth noting that his department was partly responsible for the granting of licences to make such conversions.

There was considerable confusion as to how, and if, there should be a translation between these different orders of knowledge. All modern bureaucracies weave together legal-rational charters with alternative practices drawn from habit or familiar networks, but in the Soviet context, both the rational, and its alternatives, were more extreme. To hazard a rather broad definition, legal systems enshrine normative beliefs about social convention. But when current norms are replaced by a dazzlingly bright, future Communist paradise then the nature of the law fundamentally changes, as does the role of the administration whose job is to carry out rational procedures to maintain or achieve that future condition. There are different degrees of what ought to be: absolute perfection or something achievable that hovers between the ideal and the imperfect present. The former, naturally, tallies with a regime premised on the scientific method and incontrovertible rules.

I have, of course, simplified matters for a short chapter. Not all paper representations of the world outside the bureau were quite so divorced from the observable reality as the examples given above. Officials also showed me internal documents with different figures from those publicly available. These reports, for example, showed the high percentage of seismically unsafe apartments in Almaty, an earthquake zone, about which they can do nothing since they lack the finance and, in any case, are unsure if it should be their responsibility or not since residents now 'owned' these buildings constructed by a previous incarnation of the state. But these figures and anxieties are not for public consumption.

Conclusion

The various narratives touched on here illustrate two broad modes of rationality: one is static, ideal, harmonious, the other is a scientific method that, by extrapolation, became an interventionist tool of advancement rather than a means of revealing general laws. Or to put it more simply, the former mode is genesis and/or telos; it is at once Eden and Paradise, a pre-Baconian idea of man in nature as opposed to man triumphing over nature through technology. This latter instrumental use of science exemplified the modernist Soviet enterprise but was not actually systematized until the revised 1977 Soviet Constitution was issued. Here the preface now said, 'The Soviet people, guided by the ideas of scientific communism and ... relying on the great social, economic and political gains of socialism ... formalize the principles of the USSR's social system' (Sharlett, 1978: 76). Earlier constitutions had never referred to scientific communism so overtly.

Glimpses of Eden are found in tales of the land that now lies beneath Republic Square and the increasingly densely developed southerly stretches of the city. Bees and flowers were the object of approbation here for most citizens, not architects and their creations; this natural environment made Alma-Ata marvellous rather than anything of human construction. Paradise, by contrast, is to be found in the city plans illustrating a harmonious order of classical norms and proportions with cultivated nature adorning the streets – for this is an order situated in the future. To some extent, these plans *were* realized in the centre and in the suburban apartment blocks but, in building towards this new harmony, the old order of natural fecundity was sacrificed. Temporal rhythms jarred. The earth's gradual creation of life-sustaining soil was at odds with the restless immediacy of the Soviet modernist project, which forfeited this long-term abundance in the urge to catch up and overtake the West. Eden in other words was the untempered, natural world; Paradise would be regained first by destroying and then recreating Eden. The engineer's lament for the destruction of the earth in order to provide housing might have come straight from Article 12 of the 1977 Constitution: 'The collective farms ... are obliged to use land effectively, take a solicitous attitude towards it and increase its fertility.'

A third mode of rationality has also appeared throughout this paper: its legitimating role. There is a common rhetorical association between abstract rules of physical behaviour and the sense that these rules parallel a legitimate, if not necessarily virtuous, condition: an absolute truth – which, here, is the absolute good, overriding local moralities. This appears to be the case even when such 'scientific' rules are applied to different domains. The sequence of Soviet constitutions illustrates this. Neither the 1918 constitution, nor the 1936 revision refers either to rationality or science as characteristic of communism. But, for the first time, the 1977 revision not only encoded these features but announced, moreover, that mature socialism has actually been achieved through the process of scientific communism (Sharlett, 1978: 75, 78). This, as informants in 2001 remarked slightly caustically, was something of a surprise to Soviet citizens.

Thus the third mode of rationality appears in the logical fallacy of *post hoc propter hoc*, that sequence implies causality; that what has happened is right because it has been through a rational process of historically determined laws. This, of course, is something other than the Weberian schema of rationality that looks for values, habit or telos in the driving explanation for action. The idea of the plan now appears retrospectively rather than as a forward-looking idea.

In this formulation, contingency no longer disrupts the rational process but is subsumed and contained, after the fact, into predictability. Or, to take it a step further, the bureaucratic form confers 'truth' – even if a somewhat disengaged one – upon actions and representations. Rationality becomes a rhetorical weapon in the legitimating arsenal to be appealed to, cited. This use of the idea of rationality persists. The unrestrained version of market processes enacted in Kazakhstan at the turn of the millennium was described by many as adhering to 'natural laws'. Rational bureaucratic plans no longer attempted to dam what was seen as the alternative rapacious logic of the market but rather lamely followed and formalized actual developments. But then perhaps the 'reason' of the plans always was something to think with rather than to act upon.

Notes

1. This chapter is drawn from a paper presented at the 2003 ASA conference. It was part of a panel, organized by Susanne Brandtstadter, considering James Scott's opposition of techne and metis in the context of regions where large-scale, socio-economic experiments in the name of socialism had been carried out in the twentieth century, and dispensed with in the 1990s. The regions covered in the panel ranged from China through central and eastern Europe to Kazakhstan. The central jurisdictions of all areas, however, were linked by a modernist, heightened belief in the efficacy of abstract, 'scientific' knowledge in their 'socialist' period, with a variety of responses to the collapse of this belief. A collection of papers from this panel is to be published in *Critique of Anthropology*.
2. By 'state socialism', I mean those political systems that described themselves as socialist and understood this as the necessary precursor for a desired communist state. Analytically, the term is more dubious. Following Trotsky (1970), Cliff (1974), and Binns et al. (1987) the Soviet system in particular may more accurately described as state capitalism, although this term again subsumes a number of local variations. This chapter is more concerned with the former: the self-exegetical rhetoric of the Soviet state.
3. Almaty was re-named several times. The small nineteenth-century fortress established under Nikolai II was christened Vernyi. Under the Soviets, the name changed to Alma-Ata and, after independence, to Almaty to sound more Kazakh. Here, I use Alma-Ata to denote the Soviet period and Almaty for the city since 1991.
4. This is an 'ideal type', but also one either used by Almaty's citizens to caricature the distance between officials and people, or to suggest the shortfall between this 'ideal' and the corrupt, nepotism of what actually happens.
5. Frequent, rapid swings of temperature between –35 and +35 degrees result in spalling concrete and savagely potholed roads.
6. The Virgin Lands Campaign (*Tselina*) was a scheme to plough up much of the Kazakh steppe. Ultimately, the scheme wrecked the environment since the thin soil, which could bear transhumant grazing, but no more, was particularly inappropriate for prairie-style farming.

7. I carried out fieldwork in one suburb, the main market and with a variety of urban groups.
8. A cynical view of the continuing high demands made by such regulations is that the resulting fines generated by non-compliance help to boost official salaries.

Bibliography

Beisenova, A. (1998), 'Environmental Problems in Kazakhstan', in S. Akiner, S. Tideman and J. Hay (eds), *Sustainable Development in Central Asia*, Central Asia Research Forum Series, Richmond, Surrey: Curzon Press,.

Binns, P., Cliff, T. and Harman, C. (1987), *Russia: From Workers' State to State Capitalism,* London: Bookmarks.

Cliff, T. (1974), *State Capitalism in Russia*, London: Pluto Press.

Carrère d'Encausse, H., (1994) 'The National Republics Lose Their Independence' in E. Allworth (ed.), *Central Asia: One Hundred Thirty years of Russian Dominance: A Historical Overview,* 3rd edn, Durham, NC: Duke University Press.

Ferguson, J. (1994), *The Anti-Politics Machine: 'Development', Depoliticization, and Bureaucratic Power in Lesotho*, Minneapolis: University of Minnesota Press.

French, R. (1996), *Plans, Pragmatism and People: The Legacy of Soviet Planning for Today's Cities,* Pittsburgh, PA: Pittsburgh University Press.

Harbison, R. (1991), *The Built, the Unbuilt and the Unbuildable: In Pursuit of Architectural Meaning*, Cambridge, MA: MIT Press.

Hart, K. (2005), *The Hitman's Dilemma*: *On Business Personal and Impersonal*, Chicago: Prickly Paradigm Press.

Humphrey, C. (2002), 'Remembering an "Enemy": The Bogd Khan in Twentieth-Century Mongolia', in R. Watson (ed.), *Memory, History, Opposition under State Socialism*, Santa Fe, NM: School of American Research Press.

Kurkchiyan, M. (2000), 'The Transformation of the Second Economy into the Informal Economy', in A. Ledeneva and M. Kurkchiyan (eds.), *Economic Crime in Russia*, The Hague: Kluwer.

Malyar, I. (1976), *Alma-Ata*, Alma-Ata: Zhalin.

Marx, K. (1970) 'From the Critique of Hegel's Philosophy of Right', in J. O'Malley (ed.), assisted by Richard A. Davis, *Marx: Early Political Writings*, Cambridge: Cambridge University Press.

Scott, J. (1998), *Seeing Like a State: How Certain Schemes to Improve the Human Condition Have Failed,* New Haven: Yale University Press.

Sharlet, R. (1978), *The New Soviet Constitution of 1977: Analysis and Text*, Brunswick, Ohio: King's Court Communications Inc.

Sik, E. (1992), *From the Second Economy to the Informal Economy*, Studies in Public Policy, 207, Glasgow.

—— (1994), 'From the Multicolored to the Black-and-White Economy: The Hungarian Second Economy and the Transformation', *International Journal of Urban and Regional Research*, 18(1): 46–70.

Sloterdijk, P. (1987), *Critique of Cynical Reason*, Theory and History of Literature series, vol 40, trans. M. Eldred, Minneapolis, MN: University of Minnesota Press.

Trotsky, L. (1970) *The Revolution Betrayed: What is the Soviet Union and Where is it Going?* New York: Pathfinder Press.

Verdery, K. (1996), *What Was Socialism and What Comes Next?* Princeton, New Jersey: Princeton University Press.

Wanner, K. (1998), *Burden of Dreams: History and Identity in Post-Soviet Ukraine*, Philadelphia, PA: Pennsylvania University Press.

Yurchak, A. (1997), 'The Cynical Reason of Late Socialism: Power, Pretense and the Anekdot', *Public Culture*, 9(2): 161–88.

–4–

Information Society Finnish-style, or an Anthropological View of the Modern

Eeva Berglund

Modernity is an ill-defined yet potent concept. Qualifying it by using 'post-modern', late modern or even 'neo-modern', is hardly clearer. Each of these terms summons an evolutionist rationalization of Western supremacy where moderns are ahead and the rest are behind. They also invoke science and technology. Turn-of-millennium Finland, brimming as it has been with impatient rhetoric, which extols innovation and technoscience, captures well the future orientation of modern thought and shows how it still shapes lives.

Treating modernity as an anthropological concern requires us to address technology, science and modern medicine. It is true that anthropology was born as the study of precisely those who do not have science but its empiricist stance has perhaps always predisposed it to attending to technology and science. This has been so even in the typically marginal places anthropologists have studied, where Western bias might well have obscured the presence of technoscience or minimized its significance. However, in its search for conceptual tools for understanding modern technology and science as part of culture and society, anthropology has benefited substantially from work under the late twentieth-century rubric of science and technology studies or STS. A whole range of research on the social life of science, technology and medicine – or, as I prefer, of technoscience[1] – has begun to break down the walls that kept science at a safe distance from society and politics.[2] This work has provided us with the 'theoretically informed and empirically grounded insight' (Bijker, 2003: 448) necessary for making claims about the function and significance of science in society. Anywhere that technology, science and environmental change are not ignored or bracketed out, for example in human geography and environmental history, one is likely to find reference to STS and its siblings. If we want to take seriously the consequences of modernity and to challenge the myth that 'it' is both inevitable and inevitably progressive, STS provides excellent tools. To use Bruno Latour's (1987, 1993) language, STS dismantles or refuses the Great Divide between science and belief that has been so potent a support of the divide between the West and the rest.

Influential critics of modernity elsewhere, Michel Foucault foremost, alerted us to the subtle rigidity of modern ways of creating knowledge and keeping order. These include a preference for clearly demarcated fields of expertise, for instance anthropology's own tradition of bounded field sites. I share with anthropologists of technoscience (Downey and Dumit, 1997) an impatience with boundaries, both geographical and disciplinary. On the other hand, I am alert to the dangers of ignoring limits and limitations and of generating, in Kate Soper's (2003) words, 'fuzzy constructionism'. If my text crosses borders and travels through time and skims over some details, it is in the hope that a large-scale and long-term perspective can help clarify and open up new and constructive questions.

It is based on a lifetime of intimacy with my country of birth including several years of intermittent anthropological fieldwork in Helsinki and Kainuu province.[3] It draws on cultural and economic history and on extensive documentation and analytical literature about Finland as an 'information superpower', only a fraction of which can be referenced in the bibliography. It is informed by research on Finnish environmentalism (Berglund, 2000) and guided by the preoccupations I have come upon in ethnographic and other encounters since the mid-1990s. I have noticed, in particular, how Finnish discourses construct technoscience as both inevitable and a matter of choice, a feature that STS identifies as being at the core of technoscience. Finally, I hope this account from Finland captures the historical momentum of modernity as a global phenomenon at the same time as making the case for specificity and nuance.

The Hype

Post-modern anxieties notwithstanding, science, progress and modernity are still conflated and, importantly, desired all around the world. In Finland, science and modernity have been embraced and celebrated as national achievements so that technoscience appears as mundane and taken for granted at the same time as it symbolizes a kind of specialness. By the middle of the last century science and engineering had become part of the nation's collective identity, expressed for instance in the adoption into Finnish usage of the English 'know-how'. Finland may have been a predominantly rural society until the 1960s, but it won its ticket into world society by developing its modern credentials. As many Finns say, we've been lucky that it turned out so well. Many also note that we are not simply modern; today we're at the cutting edge.[4]

Developing Finland as an information society is government strategy (see, for example, Science and Technology Policy Council of Finland, 2003) and it is exhaustively debated in the press and academic publications (for example, Castells and Himanen, 2001). Brochures and posters published by ministries, regional authorities and the Patents and Registration Office, texts ranging from policy briefings to articles in women's magazines, extol the virtues of creativity and celebrate innovative Finns. Indices of international competitiveness – whether published by the World Economic Forum, the OECD, the European Molecular Biology Organization, the Financial Times or whoever – are regularly published and discussed, particularly when Finland ranks among the top performers. The steady stream of news about the adoption of high technology by Finnish society shows no sign of abating, either in the national media or on the Internet where information on all things Finnish continues to proliferate. An on-line document posted in August 2004 states that 'approximately 2,7 million Finns or circa 70% used the Internet since the beginning of January [2004]. Women have caught up men in the Internet use ... use among thos [sic] over 50 years of age has increased the most.'[5] Another informs readers that Finland is a country 'of 5.2 million who are heavy users of media products and early adopters of media innovations.'[6]

In public representations technology, science, and progress merge into each other. As they proliferate they suggest both excitement and everydayness. Meanwhile a growing proportion of the national budget is invested in creating, transforming and transmitting scientific and technological knowledge (Science and Technology Policy Council of Finland, 2003; Pesonen and Riihinen, 2002). High, that is capital-intensive, technology accounts for a steadily increasing proportion of export revenue, and policy makers predominantly see a knowledge intensive economy as Finland's only option (Berglund, n.d.). People say that Finns trust science and experts more than other Europeans. Official documents and research suggest that the public also has considerable faith in science and technology policy (for example, Rusanen et al., 1998). The government today may be concerned about the relationship between research and society and trying to promote the public's interest in science, but the (apparent) collapse of the credibility of science and technology elsewhere in Europe (Snell, 2002: 286) has not spread to Finland, except to some extent in relation to genetic modification (Niva, 2002). In other words Finland is a curiously technophile country.

In telling this story, the role of education is usually emphasized (Castells and Himanen, 2001; Pesonen and Riihinen, 2002). A vigorous modernization drive took place in the 1960s and the number of universities was increased from six to twenty whilst government agencies and quasi-independent organizations were set up, whose aim was to enhance competitiveness through science and technology. For example, in 1967 SITRA, The Finnish National Fund for Research and Development,[7] was established. (Castells and Himanen characterize this organization as a kind of 'public capitalist' that provides funding above all for start-up companies that require high inputs of capital. It also informs policy on market needs in all high-tech sectors.)

From the 1970s, reflecting international trends, more 'research on research' (for example, on relations between science and 'the public') was commissioned, pushing policy towards an even greater emphasis on engineering skills and technological innovation. In the 1980s, management experts offered their know-how to help the cause, and there ensued a search for innovative ways to innovate. The vocabulary of NIS, or National Innovation System, gained currency in the early 1990s and is now central to government policy. According to the Finnish Science and Technology Information Service, 'Finnish science and technology policy … has been progressing from discrete scrutiny to a more comprehensive approach, in which the producers and users of knowledge are regarded as an entity. Finland is a forerunner in the development of this entity known as the national innovation system.' Although much of the literature on the economic transformation, the so-called Third Industrial Revolution, focuses on information and communications technologies or ICT, biotechnology has been an important element of science and technology policy since the mid-1990s. Here, too, there is a belief that Finnish successes exceed expectations of such a small, marginal country.

Policy is less about intervening directly in industrial policy than it is about creating the optimum environment for commercial innovations to emerge. Direct and indirect investments in research and development are currently among the highest in the OECD countries but the responsibility for being innovative and entrepreneurial is moving onto the shoulders of the public as individuals. Meanwhile communications infrastructure locks people into standard technologies.

The dangers of not pursuing an effective policy in education and research are constantly highlighted and the exhortation to strive for 'excellence' and to be 'world-class' and 'top' has become a numbing refrain. The idea of being top (*huippu* in Finnish) used to be associated primarily with athletes but is now routinely applied to entrepreneurs, scientists and engineers. The engagement between research and the rest of society, which has typically been close throughout the Nordic[8] countries (Gullestad, 1991; Miettinen and Väliverronen, 1999) is correspondingly more visible. For instance Pekka Himanen, a philosopher, has become a celebrity of sorts and parliament commissioned him to put forward his programme for fashioning the country's future as an information society. The result was a 37-page document whose title included the words 'caring', 'supportive', 'creative', and 'deep challenges' (Himanen, 2004). Himanen had already co-authored a study with Manuel Castells (2001), which analysed but also extolled the virtues of a specifically Finnish model of the information society. They compared it favourably to two other leading informational societies: relative to the US Finland is more egalitarian; compared to Singapore, much less authoritarian: a friendly and rather Nordic-looking model of the information society.

However friendly it may seem, the imperative to compete is everywhere, and the information age has made wealth creation a national duty. Meanwhile the institutions of industrial modernity – forestry, heavy industries, welfare services – have not vanished, but they are increasingly characterized as old-fashioned, complacent and rigid. The information society is contrasted with this assumed fixity and said to be about novelty and innovation. Authoritative voices support the view, as sociologists, political historians and business experts join policy and public discourse[9] in claiming that change is no longer directed from above, but emerges out of the collective fervour of hackers and other creative types. The arguments use a mellifluous language of fluidity, flexibility and flow which may be familiar to anthropologists from, among others, the work of Manuel Castells (Castells and Himanen, 2001) and Zygmunt Bauman (2003). With exemplars like these, no wonder Finnish thought has adopted fluid networks and individual creativity as both descriptive and imperative concepts.

Anthropology is well placed to attend to the human costs and cultural implications of global 'flows' (Appadurai, 1996) associated with the transition to an information-based, networked economy. Critiques of the information-society tend to focus on the question of time (for example, Eriksen, 2001) but there is a complementary route. That is, to reassess the supposed fixity of the network society's predecessor. It, too, was a changing society of noteworthy economic growth and a culture of purposeful modernity.

When all around the talk is of rupture, a reminder of continuities might be valuable in itself. More specifically, with some historical depth, Finland's amicable contract between technoscience and the public can be shown to be a far more political as well as cultural thing than the usual focus on education and research allows. I argue that technoscience was an indispensable practical element in developing industrial capitalism, but also that it had a huge if unappreciated role in forging links and commitments between the state and its citizens that Finns still cherish (Pesonen and Riihinen, 2002).

These are expressed, for instance, in the continued trust in the 'authorities' (Miettinen and Väliverronen, 1999) such as the Technical Research Centre of Finland, The National Public Health Institute and The Finnish Forest Research Institute, all established in the twentieth century in the name of the nation. If there is criticism of these authorities, it is that they are not doing their job well enough. Next I turn to why this view might be so prevalent.

Modern Finland as a Techno-political Object

Finland only became independent in 1917. Before that it had been part of the Russian Empire and before that, part of Sweden. After independence, its political and economic truths became remarkably consensual. A key consensus of the twentieth century was the country's political neutrality, geared towards safeguarding its economic and political interests while continuing to live as a neighbour of the Soviet Union. But another Finnish consensus, significant here, is that it was a forest state, its organization justified by reference to expert knowledge, notably the forest sciences.

One of the country's first prime ministers, A. K. Cajander, was not only an ardent nationalist but a key figure in establishing research and education institutions to support the forest industries (Raumolin, 1984). In 1934 he said that forest science aims at universal truths but that it is also a national science, geared towards ensuring the nation's livelihood as well as its spiritual life (Michelsen, 1995: 136). This kind of thinking supported the claim, which became even stronger and more ubiquitous after World War II, that Finland lives off the forest. This official truth in turn legitimated land use and fiscal measures like currency devaluations. It also underpinned the structure of industry and employment, not to mention geography (Lehtinen, 1991). A so-called 'forest consensus' emerged as the industry became increasingly capital- and technology intensive and forest management became more professionalized.

In the 1960s these processes led to social dislocation, but overall they were accepted as part of a national economic strategy (Donner-Amnell, 2001). It was not until the 1980s, when environmentalist concerns gained widespread media coverage, that the experts in charge of the forest sector were publicly challenged (Väliverronen, 1996). The early 1990s saw a debilitating recession, which hastened political and economic transformation. Around the time when Finland joined the European Union, in 1995, the idea of Finland as forest state was replaced by a largely Nokia-led vision of Finland as an information society. One set of experts, the forestry professions, were superseded by another set – the high-tech and innovation professionals.

The information society embodies particular economic relations and practices. After decades of high state involvement in the economy and a large public sector, in the 1990s Finland embraced the more or less neo-liberal economic principles now associated with the information age. The process, argues sociologist Pertti Alasuutari (2004), was much influenced by a desire to keep up with others, notably the other Nordic countries. But Alasuutari's use of the term ideological shift seems particularly apposite. He argues that in the grip of crises of one sort or another, in the early 1990s, the large state sector was able to reach the whole population and, ironically, persuade it of the urgent need to adjust to a pared-down public sector. Now competition and markets are as taken for granted as are the electronic noises that punctuate everyday life.

Globalization, technoscience and education, the end of the Cold War and ideology all certainly played their part in Finland's economic transformation.

Finland's makeover was also, however, a result of government by experts, something that has been brilliantly analysed by Timothy Mitchell (2002) in his studies of colonial and post-colonial Egypt.

The link between scientific and colonial projects is well made in the academic literature (Latour, 1993; Gupta, 1998) and has alerted us to considering the ongoing political as well as the technical power of technoscience – a link that Mitchell captures in the term 'techno-politics'. Yet techno-politics is everywhere, not just in neo-colonial contexts where experts rule from a distance. Furthermore, modern ways of doing things still promise improvement for millions around the world. As Mitchell (2002: 15) puts it:

> from the opening of the twentieth century to its close, the politics of national development and economic growth was a politics of techno-science, which claimed to bring together the expertise of modern engineering, technology, and social science to improve the defects of nature, to transform peasant agriculture, to repair the ills of society, and to fix the economy.

His book, *Rule of Experts* (Mitchell, 2002), is a series of detailed studies of techno-politics in Egypt. Partly inspired by STS, it follows tangled and multifarious networks that link, for instance, mosquitoes, the niceties of local etiquette and violence towards people and environments. It shows how the usual modern binaries constantly create and recreate the separation between reality and representation on which expert rule depends. The empirical studies also serve to contest the pervasive tendency of social science to grant coherence and systematicity to the apparently universal processes of modernity, capitalism or globalization. Mitchell's aim is to refuse to attribute unwarranted coherence to these forces. His account also corroborates other insights in STS: that science and technology are value laden and open to challenge; that they are central to threats to social stability (Bijker, 2003: 444). The last point in particular underlines the need to understand the interpenetration and mutual embeddedness of technoscience and politics. Mitchell's micro-studies of rule by experts carefully trace how, for example, expert authority turns stereotypes of what 'peasants' might be and do into exploitation and impoverished soils. Above all, he is alerting us to the bigger, more diffuse, politics of technoscience.

Techno-politics in Mitchell's sense, that is the imbrications of economics, knowledge practices and identity that have constituted Finland, are what I turn to now. They show that techno-politics can unfold in very different ways.

A Localized, Nuanced Modernity

Modernity in Finland was achieved under the guidance of the emergent national leaders of the nineteenth century who set in motion a process of transforming human and non-human forces under the banner of the nation. The state took responsibility for the management of the key natural resource, the forests, and set up the Forestry Board to ensure that the industry was adequately supplied with timber. It sponsored the technological and scientific expertise – especially industrial engineering and biochemistry – which transformed the timber into saleable products. By the 1920s, following the North American precedent, the state was also putting conservation into expert hands.

This institutionalization and professionalization began in the 1850s, when Finland was still an Autonomous Grand Duchy of the Russian Empire. It was in that period that the first nation-wide forest surveys were initiated. Since then Finland's forests have been regularly inventoried and records have been kept of all things pertaining to the forest.

Modern interests, like a growing awareness of culture and the spread of literacy,[10] paralleled the gradual processes that modernized relations to the forest. Raw materials were converted in ever more systematic ways into consumption goods for Russia, Europe and North America, first as sawn goods and later paper and pulp. When trade was liberalized in the second half of the century the process accelerated, leaving Finland to pursue independent economic and social policies. These opportunities invited technoscientific innovation but they also required systematic administration. No less a power than Tsar Alexander II himself encouraged the governor-general of Finland to run the administration using scientific methods. Mathematical calculation became a tool of government and forestry expertise offered itself as a method of increasing productivity as well as controlling any backward peasants still caught practising slash-and-burn agriculture.

Throughout the nineteenth century there were disagreements about what scientific, rationalized forestry should be. There were also those who did not accept the idea that the ownership of forests should be concentrated in the hands of the industry. Rather, echoing national romantic celebrations of the peasant organically tied to his territory, some argued that farmers and peasants should remain in control and ownership of the land. One prominent critic, J. V. Snellman, is quoted as saying that he hoped not too many 'scientifically educated young men will be attracted to this field, where practical work requires no scientific skills' (quoted in Michelsen, 1995: 34).

That debate took place in the 1850s. A century and many forest conflicts later, those in favour of a professionally grounded forest consensus had won (Laitakari, 1961): what was good for the forest industry was good for the nation and it was the duty of everyone to submit to the expertise of the forest profession. In return most parts of the country benefited from the jobs and comforts that the modern forest industries could provide and economic activity intensified as subsidiary industries developed. It was as part of that process that a national cult of know-how was incorporated into Finnish identity. Decades of forest consensus subsequently created an understanding of forests as national patrimony, but the connection between forests and Finnish prosperity was purposefully created and it has always contained elements of ideology and illusion.

A key feature of Finland's forest products industries has always been their reliance on export markets. Nordic imports of wood to Britain might have been a tiny fraction of its economy, but for the exporter countries Britain's need for timber between 1850 and 1900 meant a completely new intensity and expansion of production. Finland lagged somewhat behind in this process, but by the early twentieth century it was doing well in international trade.[11] Although wars were bad for the timber trade, Finland's geopolitical position made exports to both the East and the West easy and Finns were quick to exploit these opportunities. In the early years of the twentieth century, the new, capital-intensive paper and pulp industry tended to be owned by Finns even if it was not always run by Finns. Not only were experts imported, the machinery that built Finnish industry at the beginning of the twentieth century was brought in from France, England, Germany, Sweden and so on (Michelsen, 1983).

Demand for paper continued to grow and the relatively small Finnish companies created co-operative sales associations through which they approached their foreign buyers. Finland's industrialists thus quite consciously constructed a corporate economic structure where competition within the domestic setting was minimized and competitive advantage *vis-à-vis* the outside was maximized. Historian Markku Kuisma has written that co-operative traditions among paper producers going back to the 1870s 'had accustomed the leadership of enterprises to collaboration at the practical level and had, at the same time, moulded the mentality of the industrialists in a direction that emphasized not so much mutual competition as the benefits of co-operation' (Kuisma, 1993: 492, my translation). From its beginnings, the sector also entwined itself in banking and sought influence on legislation, foreign and regional policy. In turn foreign affairs were constantly reflected in the fortunes of the industry within Finland. From its beginnings, therefore, techno-politics linked Finnish enterprise to global events even as it shaped the social and biophysical landscape at home.

According to Michelsen's History of Forest Research in Finland, industrial systems of exploitation were developed prior to any scientific and technological basis for them (1995: 69) and relations with the forests on the whole were marked by pragmatic concerns. By the time of independence, scientific forestry had won the day but land remained in small-scale ownership producing a class of independent farmers (Abrahams, 1991). Productive forests have been overwhelmingly in private ownership (even today, over half of productive forest in Finland remains in the hands of small woodlot owners) and the seasonal patterns of employment created by the industry, meant that in the period from independence to about 1975, the forest industries brought substantial direct economic benefits to most households. Land reform, a state-led process of defining the boundaries of land holding and inheritance rights, was implemented in stages from the late eighteenth century to the 1860s. It created freeholders as well as landless people, but some forms of collective ownership were also institutionalized (Määttä, 1999). The majority of the population was left with some ownership rights. From the time that forest products achieved exchange value, rural households have combined agriculture with small-scale forest ownership. Most Finns no longer live on farms but an ideal of independent individualism and respect for private property is part of a cultural heritage bequeathed by the country's agrarian past (Abrahams, 1991).

As the forest industries pushed further out into more sparsely populated regions in the early years of the century, trouble brewed when local, expansive uses of forests clashed with capitalist uses. In Kainuu province for example, a kind of internal colonialism typical of resource-rich 'backwoods' was – and some say still is – the prevailing economic situation. In Kainuu, as elsewhere in the north and east, the crown, later the state, owned large areas of land, but there were also various collective forms of ownership, which came under challenge as the industry expanded (Virtanen, 2004). Stories are still told of how the men from the Forestry Board came and cheated locals by allocating poor quality mires or rocky areas to them giving the more productive lands to the companies. Revisionists now reverse the story: it was the gentlemen from town who were cheated when they were sold vast areas of effectively useless lands. The point is that independent ownership of land was fiercely and successfully defended.

The political legitimacy of the paper-and-pulp economy has since been bolstered by the fact that ownership of forests was not allowed to concentrate in the hands of a few (Donner-Amnell, 2001). Nevertheless, the forest has been, and largely continues to be, subject to the oversight of state-sanctioned experts. The aim of the national forest administration was to ensure the maximum possible sustained yield of timber. Individual forest owners, although enjoying ownership rights to their land, were thus prevented from both under-utilizing and over-utilizing their assets under a system of modern forestry that more-or-less told them what to do with their property. Official expertise has not always been well taken,[12] but until the forest conflicts of the 1990s when the sector's economic significance was already on the wane, its legitimacy was not systematically questioned.

The forest state appears, then, as a system imposed from above, modernity as the attempt to exert control from a distance. But this view must be qualified. To claim that expertise and science only developed in the Faculty of Agriculture and Forestry based in Helsinki is simply inaccurate. Secondly, although modern forestry invokes images of plantations, no Finnish forests have ever been reduced to a purveyor of mere raw material for industry. Forests are still a recreational and spiritual as well as economic resource (Berglund, n.d.) and even economically they have been exploited for all kinds of non-industrial uses. Rights to using the forest were curtailed and helped ensure that collective goals could be met well before the rise of a forest administration (Määttä, 1999).

Finnish scholars have recently become interested in the ownership of natural resources. Tapio Määttä's (1999) work on landownership shows that current Finnish thinking incorporates three core types of ownership of natural resources: non-ownable, collectively ownable and privately ownable property. Non-ownable resources include things like wild animals and protected plants (1999: 88). Collective ownership forms vary regionally with, for instance, Lapland's reindeer-herding economy having its own, region-specific regulations. Another relevant example is the ecologically valuable forests that environmentalists seek to protect, which are owned by what are effectively co-operatives of small farmers or their descendants. The layers of historically constituted ownership forms are interesting in themselves (and similar complexities exist elsewhere too, even in the UK and the US) but they are even more significant when we consider the various ways in which they are further qualified and specified. Private property is always subject to the requirements of the broader common good.

So although land reform encoded the rights of some relative to others, it did not create many literal enclosures. It also meant that even private ownership remained subject to the needs of others. These are varied and include rights to temporary use like the so-called everyman's right to another's land to collect mushrooms or berries, but also zoning based on the needs of urban centres or technological interventions such as large-scale irrigation. These kinds of qualifications on ownership do not challenge the legal principle that only one subject can have ownership rights over one object; what they do challenge is the idea that ownership of a thing could mean absolute power over it (Määttä, 1999, section 3.3).

Thus Finnish modernity was never quite as centralized or as capitalized as the Anglocentric idea of industry-led enclosure implies. This may help explain why modernity has so easily been embraced as a dimension of collective identity and how technological and expert-led forest use became part of national identity. At the same time this identity was also able to accommodate a self-image of backwoods men and women adapting to and learning from adverse conditions, whether local famine or global recession.

Finnish experiences echo those in the other Nordic countries. Orvar Löfgren (1980) notes that compared to their English or French counterparts, Scandinavia (including Finland) maintained more complex forms of collective ownership and correspondingly more corporate forms of social organization. He also notes that when it did so the modern state made itself felt via the activities of local administrators. Many were undoubtedly, as he notes, zealous bureaucrats keen to keep good records, but many were also farmers and not a distant, locally unaccountable class making unreasonable demands.

It is possible to say that the rule of experts in Finland served the interests of a majority at the same time as it advanced the fortunes of capitalists and narrow elites. Thus Mitchell's idea of techno-politics as a fixing of standards emanating from a far-away centre (or far-away centres) may be overly totalizing as well as pessimistic. Let me invoke an interview at a small biotechnology lab in rural Finland, where we were discussing the workers' relationship as university employees to society at large and specifically to the state. Two women were emphatic in exclaiming 'we are the state!' I was taken aback by their insistence and clarity on this point but it echoes a close identification of the state with the welfare of the public that one frequently comes across and which, with some variation, is characteristic of other Nordic countries too (Gullestad, 1991). The link was consciously made via the idea that society in the form of the state provides the individual with an education and gives him or her the same chance as everyone else at success. Universal education can be interpreted as co-optation by the state and it has undeniably operated as ideology and obscured social inequalities but it also provides opportunity and promises some security. Indeed, the close link with the state and with the welfare it provides is now increasingly viewed as something that needs to be consciously nurtured lest it be taken away. Maybe identification with the state needs to be asserted because it is no longer as straightforward as it once seemed.

To conceptualize the current situation one might say that the imperative of national economic growth made Finland into a resolutely modern place, if not managed as a uniform system then at least guided by principles of universality. It is also resolutely modern in that these aspirations influenced the development of a national culture where identity was constructed out of symbolic as well as practical preoccupations with land or territory. By the second half of the twentieth century, to be Finnish was to identify with forest and with nature but it was also, I suggest, to be modern, and that meant being open to change.

Modernity meant commitment to open-endedness even in the monolithic-looking forest sector where the best forest science has always been a matter for debate. Thus a charitable interpretation of the image of a recalcitrant forest expertocracy might be that it was a posture born out of political expediency. When in the 1990s the forest sector faced serious challenges from environmentalists, the latter based their criticism of national techno-politics on the observation that their science was newer and better science (Berglund, 2000). As the industry responded, the process was far removed from the media image of a simple duel (Väliverronen, 1996) as throughout the conflicts everyone emphasized the imperative of progress and their own superior scientific understanding. It turned out that debating Finland's forests was to talk about society. Expertise itself was judged on explicitly social and political criteria and it became possible to demand that the problem itself be reframed: not how to make most profit out of the land but, rather, whether the land could survive the demands put on it. Yet underlying even this kind of survivalist rhetoric has been the interest of the nation represented by the state. As unemployment, trade, economic uncertainty and regional disparities face decision makers at all levels, science as cutting edge gives way to science as pragmatic, science as good for the people or science as good for the economy. The economic implications of a forestry station or a biotechnology laboratory count for more in public debate than does abstract scientific merit.

In sum I have argued that expertise, specialization, know how, professionalism but also open-endedness, were adopted as practices, but importantly also as explicit aspects of Finnish national identity. As a principle of arranging social relations and cultural norms, technoscientific thought still has a huge hold on the Finnish imagination. This is not because it was imposed, not because innovation is indigenous to Finns, nor even because technical expertise operated in the interests of most Finns, but above all because it has formed a significant part of collective identity and practices that provide shared orientation. Today that requires putting faith in the NIS, Finland's National Innovation System.

In embracing this self-image, and in developing its forestry infrastructure, Finnish efforts were significantly helped by the legacy of corporate economic management typical, as Löfgren noted, of all Nordic countries. Finland, with its co-operatives of capitalists, its peasant individualists and its qualifications on private property, became a viable trading partner, possibly because of its not-quite-modern regimes of ownership.

Modernity – Some Final Thoughts

Finland's information economy is still finding its shape and it is hard sometimes not to be overwhelmed by the abundance of its main product, data. Incremental addition of detail and shifts of emphasis notwithstanding, the NIS is now as taken for granted as Finnish excellence. But a certain panic that the good times are about to end is equally evident. Furthermore, rewriting this chapter in 2005 I note that more critical tones have begun to appear in the debate: is the price of success too high?

This dovetails with the anthropological thread I have wanted to weave into my account and now want to bring to the fore – the effects of modernity and non-modernity. For as we scan the planet for transformations and innovations, it is clear that some creative knowledge practices partake of modernity and are thus a ticket to more of the good life, whereas others are dismissed as impurities or problems needing to be fixed.

Three very different examples come to mind. The first is described by Christine Hine (2000). During fieldwork on genetics research she regularly came across situations where a standard laboratory procedure inexplicably stopped working. Generally the reactions of the researchers amounted to shoulder shrugs: some things apparently just don't have a rational explanation. Because of her own expertise in computing, she had a particularly interesting experience when somebody's computer would not let the user log on. Hine offered to apply her expertise to the problem, but the researcher cut her off with 'It is OK, it is playing up, it does that with mine from time to time for some reason.' Hine (2000: 68) concluded that for this person the computer was not a 'logical, knowable machine whose problems required rational explanation.' In such situations apparently incommensurate styles of thought can co-exist without need for comment. Nobody would think to question the scientist's modern credentials.

Another example might be the way designers, engineers and architects in Finland still say they turn to nature for inspiration, in both technique and aesthetics. The Finnish modernist architect Alvar Aalto was an outspoken advocate of both regional context or tradition and universal modernism in architecture (Pallasmaa, 2005: 269). His work combined modern and vernacular or non-modern features drawing from 'ageless peasant traditions' according to Pallasmaa (2005: 277) – and was consistently praised for it.

But mixing and matching is not always acceptable. Nor does experimentation with unfamiliar techniques always signal creativity or innovativeness. It can even signal failure. I am referring to those people and places whose actions were long ago imagined as pre-modern and that still struggle to claim a share of modernity's benefits. The terms 'indigenous' and 'indigeneity' have replaced 'primitive' as the opposite of modern but they articulate a similar value judgement about otherness and incommensurability as described in Akhil Gupta's *Postcolonial Developments* (1998). In this analysis of farming practices in Uttar Pradesh, Gupta describes how farmers regularly blend humoral methods – that is, ones based in assessing the appropriate wetness of fields – with scientific farming methods. The capacity to make creative use of apparently incommensurate knowledge systems, of humoral as well as scientific knowledge, actually devalues Indian farmers in the eyes of the powerful. More specifically, they are seen as unfit for inclusion into pure categories of identity. Neither pure savage nor proper modern, they claim neither of the two identities offered to them: of privileged holders of ecological wisdom, or shining examples of modern development. In the schema operationalized in decisions of the World Bank, for instance, they have 'failed' to be either properly modern or properly indigenous. Yet the farmers Gupta describes are technologically creative. Furthermore, they disrupt the political schema of morally laden identities such as modern or indigenous.

Juxtaposing these examples highlights two core preoccupations of anthropology: taxonomies and identity. The question of identity may appear dull, even over-researched, but the reality is that that identity and its classification remains hugely important. We must pay attention to it. The attribution of identity continues to inform how the rewards and costs of globalization are distributed and the difference between marked and unmarked, desirable and problematic, is still most often defined by reference to some assumed neutral or standard version of the modern. Finnish identity as a modern identity is so unquestioned that boundary crossings invite little or no scrutiny. The Finnish public and the Finnish state mix and match nature and culture, modern and pre-modern, but remain, for all that, resolutely modern and are even treated as the vanguard of a new, ever more innovative modernity. In short there are consequences for the way Finns identify themselves and, even more importantly, are recognized as modern.

Perhaps there is scope for learning from as well as critiquing Finland's recent experiences. And if anthropological aims are not limited to the productivist ones that inform the academy, neither are they defined by the discipline's tendency to go (uncritically) native. Anthropology can maintain the militant middle ground, to borrow Michael Herzfeld's (2001) phrase, from where moralizing may be avoided and from where one can simply assert that techno-politics unfolds in complex tensions. Reminding an academic audience of complexity is not, however, enough. There is scope, maybe even a need, for a broader discussion about why not having become fully modern remains a key obstacle when for others it is a key resource. Considering modernity as an identity is a start, but the usefulness of this approach is, I think, considerably enhanced if we do not sever identity from the material and the technical forces that accompany it.

Notes

1. Modern science and technology need each other (Latour, 1987). The key function of the distinction between them is to create illusions, firstly that science and society are separate and secondly, that science is independent of or prior to technology.
2. Prominent analytical aids for anthropologists have included actor networks and hybrids of nature and culture (Latour, 1993). STS has also developed ways of understanding technical arrangements as forms of political order (Winner, 1980), has pointed out continuities between conceptual and literal violence (Nandy, 1988; Haraway, 1997) and demonstrated the political power of what Sharon Traweek (1988) has aptly called scientific cultures of no culture. Shapin and Schaffer's (1985) influential account of Hobbes' and Boyle's seventeenth-century disagreement over the air-pump is a sophisticated analysis of how the boundary between science and politics developed as it did. The influences between this historical work and studies of contemporary laboratory science go both ways.

3. I acknowledge the financial assistance over the years of the Wenner-Gren Foundation (individual research grant 6963), the Joint Committee of the Nordic Research Councils and The British Academy. Many thanks to the three editors, to Karen Armstrong and to Sakari Virtanen for constructive feedback.
4. The political history of Finland by Pesonen and Riihinen (2002) examines recent transformations in historical context but also exemplifies the congratulatory tone in which recent events tend to be discussed.
5. From e-finland Web site, which describes itself thus: 'e.Finland.fi is built and maintained in wide cooperation with several ministries and national promoters of information society. The basic idea of the site is to provide all interested parties a single site offering a concentration of Finland's information society activities, thus eliminating the need to search for information from several sources' The quotation in the body of the text is from http://e.finland.fi/netcomm/news/showarticle.asp?intNWSAID=26947, accessed December 2005.
6. The Virtual Finland Web site is produced by the Ministry for Foreign Affairs. http://virtual.finland.fi/, accessed December 2005.
7. It is hard not to pause at the Finnish name of this organization: despite its official translation, a more accurate one would be Commemoration Fund of Finnish Independence.
8. Norway, Sweden, Denmark, Iceland and Finland.
9. The literature is enormous, put out by the academic and the so-called 'grey' press as well as in policy briefings. Much of it is reads as a celebration as much as an analysis of Finland's recent past. See, for example, Häikiö (2001) or Pesonen and Riihinen (2002).
10. In English, see Pesonen and Riihinen (2002).
11. I am indebted to Sakari Virtanen for teaching me about the history of the forest industries. Socio-historical discussion and references in English can be found in Lehtinen (1991), Berglund (2000) and Donner-Amnell (2001).
12. An important point made in conversation by, among others Sakari Virtanen and Tim Ingold.

Bibliography

Abrahams, R. (1991), *A Place of Their Own: Family Farming in Eastern Finland,* Cambridge: Cambridge University Press.

Alasuutari, P. (2004), 'Suunnittelutaloudesta kilpailutalouteen. Miten muutos oli ideologisesti mahdollinen?', *Yhteiskuntapolitiikka* 69(1): 3–16.

Appadurai, A. (1996), *Modernity at Large: Cultural Dimensions of Globalization,* Minneapolis: University of Minnesota Press.

Bauman, Z. (2003), *Liquid Love: On the Frailty of Human Bonds,* Cambridge: Polity Press.

Berglund, E. (2000), 'From Iron Curtain to Timber-Belt: Territory and Materiality at the Finnish-Russian Border', *Ethnologia Europea* 30(2): 23–34.

—— (n.d.), 'Forests, Flows and Identities in Finland's Information Society', manuscript under review for special issue of *Cultural Studies*, edited by Phaedra Pezullo.

Bijker, W. E. (2003), 'The Need for Public Intellectuals: A Space for STS' (Pre-presidential Address, 2001 Annual Meeting of the Society for Social Studies of Science), *Science, Technology and Human Values*, 28(4): 443–50.

Castells, M. and Himanen, P. (2001), *Suomen Yhteiskuntamalli* (simultaneously published in English as *The Finnish Model of the Information Society*), Helsinki: WSOY and SITRA (Finnish National Fund for Research and Development).

Donner-Amnell, J. (2001), 'To Be or Not to Be Nordic? How Internationalization Has Affected the Character of the Nordic Forest Industry and Forest Utilization in the Nordic Countries', *Nordisk Samhällsgeografisk Tidskrift*, 33: 87–124.

Downey, G. L. and Dumit, J. (1997), 'Locating and Intervening: An Introduction', in G. L. Downey and J. Dumit (eds), *Cyborgs and Citadels: Anthropological Interventions in Emerging Sciences and Technologies,* Santa Fe, NM: School of American Research Press.

Eriksen, T. H. (2001), *Tyranny of the Moment: Fast and Slow Time in the Information Age,* London: Pluto Press.

Finnish Science and Technology Information Service, http://www.research.fi. Accessed December 2005.

Gullestad, M. (1991), 'The Transformation of the Norwegian Notion of Everyday Life', *American Ethnologist*, 18(3): 480–99.

Gupta, A. (1998), *Postcolonial Developments: Agriculture in the Making of Modern India,* Durham, NC: Duke University Press.

Häikiö, M. (2002), *Nokia: The Inside Story,* London: Pearson Education.

Haraway, D. (1997), *Modest_Witness@Second_Millenium:FemaleMan _Meets_ OncoMouse – Feminism and Technoscience*, London and New York: Routledge.

Herzfeld, M. (2001), *Anthropology: Theoretical Practice in Culture and Society*, Oxford: Blackwell.

Himanen, P. (2004), *Välittävä, kannustava ja luova Suomi: Katsaus tietoyhteiskuntamme syviin haasteisiin,* Eduskunnan Kanslian Julkaisu 4. Helsinki: Eduskunnan Kanslia.

Hine, C. (2000), *Virtual Ethnography,* London: Sage.

Kuisma, M. (1993), *Metsäteollisuuden Maa*, Helsinki: Suomen Historiallinen Seura.

Laitakari, E. (1961), 'A Century of Finnish State Forestry, 1859–1959', *Silva Fennica*, No.112, Helsinki.

Latour, B. (1987), *Science in Action*, Cambridge, MA.: Harvard University Press.

—— (1993), *We Have Never Been Modern,* New York and London: Harvester Wheatsheaf.

Lehtinen, A. (1991) 'Northern Natures: A Study of the Forest Question Emerging Within the Timber-Line Conflict in Finland', *Fennia*, 169(1): 57–169.

Löfgren, O. (1980), 'Historical Perspectives on Scandinavian Peasantries', *Annual Review of Anthropology,* 9: 187–215.

Määttä, T. (1999), *Maanomistusoikeus: Tutkimus omistusoikeusparadigmoista maaomaisuuden käytön ympäristöoikeudellisen sääntelyn näkökulmasta*, Helsinki: Suomalainen Lakimiesyhdistys.

Michelsen, K. E. (1984), 'Paperiteollisuutemme raaka-aineen valmistus aluksi tuontitaidon varassa', *Tekniikan Waiheita*, 4: 6–9.

—— (1995), *History of Forest Research in Finland. Part 1: The Unknown Forest*, Helsinki: The Finnish Forest Research Institute.

Miettinen, R. and Väliverronen, E. (1999), 'In Science and Technology We Trust: On Public Understanding of Science in Finland', in R. Miettinen (ed.), *Biotechnology and Public Understanding of Science*, Helsinki: Publications of the Academy of Finland.

Mitchell, T. (2002), *Rule of Experts: Egypt, Techno-politics, Modernity*, Berkeley: University of California Press.

Nandy, A. (1988), *Science, Hegemony and Violence*, Delhi: Oxford University Press.

Niva, M. (2002), 'Gene Technology in Food Production and Consumer Interpretation of Risk', *Wissenschaftlicher Beitrag*, 50(3): 122–9

Pallasmaa, J. (2005), *Encounters. Architectural Essays*, Helsinki: Rakennustieto.

Pesonen, P. and O. Riihinen, (2002), *Dynamic Finland: The Political System and the Welfare State*, Helsinki: Finnish Literature Society/Suomen Kirjallisuuden Seura.

Raumolin, J. (1984), 'The Formation of the Sustained Yield Forestry System in Finland', in H. K. Steen (ed.), *History of Sustained-Yield Forestry: A Symposium, 1983*.

Rusanen, T., Von Wright, A. and Rusanen, M. (1998), 'Finland' in J. Durant, M. W. Bauer, and G. Gaskell (eds), *Biotechnology in the Public Sphere: A European Sourcebook*, London: The Science Museum.

Science and Technology Policy Council of Finland (2003), *Knowledge, Innovation and Internationalization*, Sixth Triennial Review, Helsinki: Science and Technology Council of Finland.

Shapin, S. and Schaffer, S. (1985), *Leviathan and the Air-Pump: Hobbes, Boyle, and the Experimental Life*, Princeton: Princeton University Press.

Snell, K. (2002), 'Biotekniikkapolitiikan kansalaiskuva: kansalaiset, kuluttajat ja ihmiset Suomessa ja Euroopan Unionissa', *Sosiologia*, 4: 285–95.

Traweek, S. (1988), *Beamtimes and Lifetimes*, Cambridge, MA: Harvard University Press.

Väliverronen, E. (1996), *Ympäristöuhkan Anatomia: Tiede, Mediat ja Metsän Sairaskertomus*, Helsinki: Vastapaino.

Virtanen, S. (2004) *Siniset Metsät, Vihreä Kulta: Kainuun Metsätalouden Historia*, Kajaani: Oulujoen Uittoyhdistys.

–5–

Nga Rakau a te Pakeha

Reconsidering Maori Anthropology

Amiria Henare

Despite the wide geographical dispersal of sites from which social anthropology draws its ethnographic authority, conventional histories of the discipline have emphasized the contributions of metropolitan academics over those of scholars who conducted their analyses – not just their fieldwork – in places outside America and Europe (e.g., Kuper, 1996; Kucklick, 1993; Stocking, 1992). Such a focus not only obscures the development of schools of thought and anthropological practice distinct from those that emerged in Cambridge, London or Chicago, but also the collegial networks of a discipline whose practitioners travelled widely and corresponded frequently – relationships that were perhaps closer in the early twentieth century than they are today, notwithstanding the advent of electronic communications. Yet in providing an account of how anthropology unfolded in New Zealand, with an emphasis on Maori participation in the discipline, this chapter does not merely offer a historiographic corrective. The intention is to draw out the effects this image of a Euro-American centre for disciplinary practice has had on the theoretical orientations of 'mainstream' social anthropology – the ways it has recursively influenced the development of anthropological theory.

A primary motivation for this discussion is that, in New Zealand, despite a long and internationally renowned history, social anthropology by and about Maori people is today virtually a thing of the past. Whereas local departments of anthropology once specialized in Maori and Pacific ethnography and many Maori trained in the discipline prior to embarking on academic careers, today there does not appear to be a single Maori scholar employed in any of the country's six anthropology departments (despite burgeoning numbers of Maori academics in other disciplines), nor anyone who lists among their research interests the study of contemporary Maori culture. There are obvious reasons for this – the establishment of Maori Studies departments from the 1970s, which became a way out from under the disciplinary umbrella of anthropology for Maori scholars and specialists – as well as the diversions offered by political activism and the opportunities, which opened for a variety of 'cultural experts', including consultancy positions, research work and offices in tribal bureaucracies. Yet, in explaining these moves, scholars have often accentuated such political economic factors over the impact of changing theoretical preoccupations within anthropological discourse itself. I consider how changes *within* the discipline might have contributed to the decline of Maori scholarly participation, and what the significance of these shifts might be for anthropology at large. I suggest that distinctive approaches to anthropology arose in New Zealand which were influenced by Maori conceptions, but have not found favour in the 'mainstream' community of scholars that defined the parameters of the discipline. This divergence is attributed to manifestations of the reflexive turn in anthropology, which, far from democratizing the discipline and opening it up to diverse subjectivities, have worked to reinforce old assumptions about the nature of culture, the dynamics of cultural change, and the authority of anthropological knowledge.

Maori in Anthropology

The beginnings of Maori participation in what would become anthropology, not only as subjects but as analysts of their own culture, may be traced back to the early nineteenth century. At this time English missionaries arrived in New Zealand, questioning Maori about their beliefs and practices with a view to religious conversion. Soon after, colonial administrators and collectors like Governor George Grey used techniques and theories from the emergent discipline of ethnology in their implementation of imperial power. Grey recruited Maori informants to assist him in learning local concepts in order to 'successfully govern ... a numerous and turbulent people' (1885: vii–xi). Described by Stocking (1987: 81) as 'one of the more perceptive ethnographers of his day, and author of some of the most influential ethnographic work of the [nineteenth] century', Grey was tutored by Maori chiefs and *tohunga* (priestly experts), among them Wiremu Maihi Te Rangikaheke of the Ngati Rangiwewehi tribe, who produced some 800 pages of original manuscripts on Maori cosmologies and tribal histories and regarded himself as Grey's mentor and friend (Curnow 2005).

Later in the nineteenth century, similar relationships developed between Elsdon Best, a pioneer of Maori ethnography, and Maori elders including Tamati Ranapiri, a learned man of the Ngati Raukawa tribe. Like Te Rangikaheke to Grey before him, Ranapiri wrote letters in Maori in response to Best's ethnographic enquiries, explaining key Maori concepts, most famously that of *hau* or the breath of life, later described by Marcel Mauss as 'the spirit of the gift'. (Best's quotations of Ranapiri's letters were picked up by Mauss and developed into his seminal theory of reciprocity and gift exchange.) Later, as a professional ethnologist at New Zealand's national museum, Best drew extensively on his correspondence with

Ranapiri and other Maori in producing his corpus of ethnographic writings, often (but not always) acknowledging his sources.

Much of Best's work was printed in the *Journal of the Polynesian Society,* first issued in 1892, along with numerous accounts of Maori history, cosmology and ritual practice written by colonial scholars and by *kaumatua* (Maori elders). Men such as Timi Waata Rimini of Ngati Awa, Wiremu Te Kahui Kararehe of Taranaki and Hare Hongi of Ngapuhi (who also went by the name of H. M. Stowell) contributed manuscripts in Maori on topics such as early Polynesian migrations to New Zealand and ancient chants. These were translated (with varying degrees of intervention) by the *Journal*'s first editor, Stephenson Percy Smith, and published with commentaries from Best, Smith, and other amateur ethnologists, as well as from Maori experts in tribal lore and history. Indeed, the bulk of all Maori material published in the *Journal of the Polynesian Society*, at least in its first years of operation, was based substantially on material written or recounted by Maori elders about their own people (Sorrenson, 1992: 34).

This early phase of proto-anthropological discussion in New Zealand was characterized by enduring and often close relationships between European (mainly British) analysts and their knowledgeable Maori informants. The scholars who published ethnological observations about Maori at this time were often colonial agents who used language redolent with imperial prejudice but this did not preclude them from developing life-long associations with men whose learning in the *whare wananga*, the schools of Maori learning, they grew to respect. Many of these ethnologists came to take seriously Maori ideas about the nature and workings of *taha Maori,* grouping these for their own convenience under headings such as 'cosmology', 'origin myths' and 'spiritual conceptions'. Some found their own most fundamental beliefs challenged by those of their informants and came adrift, at least for a time, from their ontological moorings.[1] Others glossed what they were told as misguided 'superstitions' and primitive understandings, thus shielding their own beliefs from question. Echoes of such attitudes can still be found in anthropological scholarship today, as I argue, in discussions of the 'invention' of Maori tradition.

Maori Concepts and Maori Anthropology

Over the past two decades, a number of publications have appeared that present contemporary 'Maori culture' variously as a product of naïve or strategic essentialism on the part of Maori and as an artefact of anthropological scholarship (for example, Hanson, 1989; Webster, 1998; Kolig, 2002; cf. Kuper, 2003). Though diverse in their theoretical orientations, these analyses appear to share a view that, since colonization, Maori have progressively lost that which made them distinct from their colonizers, becoming assimilated in fact – if not in their own estimation – to liberal democratic modernity and global capitalism. Instead of a distinctive way of being that was organically (and unreflexively) lived, it is at least implied, contemporary Maori (and many who write about them) hold an image of Maoriness that bears little resemblance to that which pertained prior to European contact. Though each author has different views about the implications of this predicament – some suggesting that it is typical of all modern cultures, others singling Maori out as particularly ripe for critique – they seem to agree that what passes for 'Maoriness' these days arises from the mingling of at least two (culturally) different approaches to the world, and that contemporary Maori 'culture' is in many ways a discursive construct.[2]

Other writing has queried the epistemological orientations of such scholarship, noting the way in which it resolves (cultural) difference by recourse to different 'knowledges' or 'epistemes' overlaying a fundamental (and, it is implied, universally recognized) humanity (Strathern, 1990; Viveiros de Castro, 1998, 2002; Henare, Holbraad and Wastell, 2007). Certainly, the conceptual alterity that became evident to early scholars in New Zealand in the process of comparing themselves with Maori (and vice versa) has routinely been interpreted by anthropologists and others as evidence of differing 'knowledge systems' that are held to have merged in the course of colonial history. Recent critiques have traced such interpretations to (Euro-American) assumptions about the nature of reality and to the second-order status accorded the concept of 'knowledge'. In the philosophical universe that continues to support the foundations of much anthropology, a disjuncture is still upheld between that which 'is known', and that which simply 'is', despite anthropologists' best efforts at reflexivity (Viveiros de Castro, 1998).

What 'is' in this particular cosmology (which of course also constitutes the philosophical baseline of the natural sciences) is that which is held in common – our shared humanity, our shared universe, 'the' world – and difference in this scheme is rendered as a product of our various knowledges *about* it. Viveiros de Castro (1998) has characterized this position as 'multiculturalist', arguing that such a view is widespread in contemporary anthropology. In contrasting it with the 'multi-naturalist' ontologies of his Amazonian informants, in which a multiplicity of 'worlds' obtain according to the kind of body one inhabits, he demonstrates that assumptions about the things that seem most self-evident cannot be taken for granted. The point of such a critique is not that it is wrong for anthropologists (or indeed scientists) to assume a shared humanity or universe, but that we might think twice before claiming that our informants have similar conceptions.

If it is unwise to assume that Maori people in general share the same ontological assumptions as the anthropologists who represent them, might it not be just as presumptuous to imply that all anthropologists draw upon the same conceptual repertoire, one whose origins lie exclusively in America and Europe (cf. Sahlins, 1999: 411–12)? My argument is that distinctively Maori concepts endure and are evident in the long-standing tradition of Maori anthropology in New Zealand.

That early Maori and their European associates did not share the same assumptions – at least to start with – was evident to anyone who engaged in proto-anthropological discussions in early New Zealand. British subjects sailing from post-Enlightenment Europe to nineteenth-century New Zealand encountered people who occupied permeable, generative 'worlds' inhabited by *taniwha* (which the Europeans described as 'spirit-guardians'), *patupaiarehe* ('supernatural beings') and animal, plant and human *atua* ('ancestor gods'). The mundane activities of *tangata* or people took place in Te Ao Marama, the world of light and life, while the power of the ancestor gods originated in Te Po, the world of darkness, death, and superhuman beings. Eruptions of ancestral power into Te Ao Marama were marked by *tapu*, a force dangerous to ordinary people and surrounded by prohibitions; while everyday life and women in general were *noa* or free from restriction. People, birds, rocks and other things all had *tinana* (bodies), *wairua* ('spirit'), *mauri* ('life force'), and *hau* (vitality, literally 'breath'). When Europeans arrived, they were first taken to be creatures from Te Po, *tupua* ('goblins') or *patupaiarehe*, and then *pakeha*, people with unheard-of ways of being. In the process, new ontological zones emerged – Te Ao Maori, the world of normality and *maori* or ordinary people, alongside Te Ao Pakeha, the world of Europeans, marked by alterity. As many more Europeans came to New Zealand, eventually far outnumbering the country's original settlers, Maori learned to move back and forth between these *ao* or 'worlds', adopting new technologies and new concepts in order to continue living. During this time, other *ao* emerged – *te ao tawhito*, the ancestral, ancient world; *te ao hurihuri*, the changing world; and *te ao hou*, the new world of the younger generation, linked by *whakapapa* ('genealogy') and other enduring conceptions. In the early 1900s, these shifts and continuities were memorialized in a *whakatauki* or proverb by the Maori leader, politician and sometime anthropologist Apirana Ngata, exhorting young Maori to learn new ways while holding fast to the teachings of their ancestors:

E tipu e rea i nga ra o to ao
Grow, child, in the days of your world
Ko to ringaringa ki nga rakau a te pakeha
Your hand to the weapons of the pakeha
Hei oranga mo to tinana
As an existence for your body
Ko to ngakau ki nga taonga a o tipuna
Your heart to the treasures of your ancestors
Hei tikitiki mo to mahuna
As a topknot for your head

Ko to wairua ki te Atua
Your spirit to Almighty God
Nana nei nga mea katoa
Who is the giver of all things!

Ngata urged young Maori to take up the 'weapons' of Europeans, including disciplines such as anthropology, in order to ensure the physical survival of their people while holding fast to te Atua – God in the singular – in their *wairua* or 'spirit'. At the same time, their *ngakau* ('mind-heart') would uphold the teachings of their ancestors, bringing those 'worlds' with them into the present.

The Dominion Museum Expeditions 1919–23

In the early 1900s, Ngata enacted this philosophy by supporting a series of state-of-the-art ethnographic field expeditions to study Maori life in different parts of New Zealand's North Island. These trips were inspired by accounts of the 1898 Cambridge Expedition to the Torres Strait, and were directly encouraged by W. H. R. Rivers of Cambridge during his visit to Wellington in 1915.[3] The expeditions produced wax cylinder sound recordings and nitrate films of both traditional and contemporary aspects of Maori culture, as well as written records and collections of artefacts, and were undertaken by a team of Maori and *pakeha*[4] ethnographers led by museum ethnologist Elsdon Best, based at the national museum. No university posts yet existed in the fledgling discipline of anthropology in New Zealand but ties had been established with professional scholars in Britain, including Rivers and his colleague A. C. Haddon, who alerted the New Zealanders to developments in British anthropological thinking while including publications like the *Journal of the Polynesian Society* among their own readings. The expeditions were conceived in the climate of this scholarly exchange, while serving a Maori political agenda to ensure the persistence of old skills and knowledge among Maori. They set a precedent for a form of anthropology that sought to ensure the persistence of old concepts and techniques. Motivated by more than a 'salvage' mentality, the Maori scholars on the expeditions (aided by their *pakeha* colleagues) planned to deploy this material in ensuring continuities between the past, present and future of Maori people.

The core expedition team for the field trips was made up of Best, the ethnologist, James McDonald – the museum's photographer, film-maker and draftsman – and Johannes Andersen of the Turnbull Library, a keen amateur ethnologist with a particular passion for music and string-games. The team was regularly joined by Peter Buck (also known as Te Rangi Hiroa) – a Maori medical doctor, scholar, and politician, who went on to play a prominent international role in Pacific anthropology as director of Hawaii's Bishop Museum and as a visiting professor at Yale. Together with Apirana Ngata, Buck had been a student at Te Aute College, the famous Maori boys' high school, and was also a member of the Polynesian Society and the Young Maori Party (formerly the Te Aute Old Boys Association), an organization dedicated to Maori cultural and economic renewal. Buck was elected to parliament in 1909, holding his seat until 1914. When war broke out, he and other Maori MPs helped recruit a volunteer Maori contingent, which he joined as medical officer. After fighting at Gallipoli, Buck was promoted to second-in-command of the Maori Battalion. Posted to hospital duty in Britain, he met Sir Arthur Keith of the London Hunterian Museum and the eugenicist Karl Pearson, both of whom encouraged his interest in physical anthropology, loaning him instruments to measure the men under his command (Luomala, 1952: 39). Buck returned to New Zealand with his battalion in 1919, where he participated in the Hui Aroha, a gathering of love and mourning organized by Ngata at Gisborne to welcome the soldiers (and the spirits of their dead comrades) home from the war. This event provided the occasion for the first of four ethnographic expeditions funded by the Dominion Museum.

To record the Hui Aroha and associated traditional cultural activities, Best and McDonald travelled north to the East Coast with Andersen. Ngata, MP for Eastern Maori, supported the expedition (Craig, 1964: 185), arranging for Te Raumoa Balneavis, the Native Minister's secretary, to set up contacts among the local people and to facilitate the gathering of ethnographic data (*Annual Report of the Dominion Museum*, 1919: 29; Dennis, 1996: 292). The team collected a wealth of material, including film footage and photographs of string games, traditional dances and cooking techniques (*Annual Report of the Dominion Museum*, 1919: 28–9).[5]

The success of the Gisborne experience encouraged McDonald and Best to plan a second expedition, to the tourist centre of Rotorua in 1920. Here they attended a gathering organized to welcome the Prince of Wales (later George VIII) to New Zealand. Ngata, an enthusiastic reader of anthropological writings, again encouraged the trip, keen for Best, Andersen and McDonald to 'continue their researches' (Gibbons, 1992: 188). The party was billeted at the Rotorua racecourse by Best's contacts among the Tuhoe tribe (Craig, 1964: 186), where McDonald filmed preparations for the welcoming ceremony, as well as activities performed for the camera like old techniques of stone-drilling, dart-throwing, flute-playing, the making of fire by friction, hand games and string figures. The film, restored and edited in 1986, also shows the *powhiri* or welcoming ceremony for the Prince.[6]

In 1921 a third Dominion Museum expedition set out down the Whanganui River, where the team spent several weeks collecting artefacts and records of traditional skills and knowledge at the settlements of Koriniti (Corinth), Hiruharama (Jerusalem) and Pipiriki. They were joined at Koriniti by Buck, eager to apply in the field the anthropometric techniques he had picked up in Britain. Buck also developed a personal interest in Maori textile manufacture (Condliffe, 1971: 146), while McDonald shot some 5,500 feet of film depicting skipping games, flax weaving, the making and setting of eel pots and divinatory rites, and took hundreds of still photographs.[7] At Koriniti he arranged the purchase from Rihipeti Aperaniko of artefacts for display at the 1924 London Empire Exhibition.[8] The team also received presentations from local people for the Museum, including 'a very fine canoe baler', an eel basket, a stone adze, kits, 'food baskets of many kinds, belts etc.' Upon their return to Wellington, these artefacts were placed on display in the Museum, and Best thanked the local people for their enthusiastic assistance in the Museum's *Annual Report* (*Annual Report of the Dominion Museum,* 1921: 19).

The fourth and final expedition of the Dominion Museum ethnographers was in 1923, when Ngata invited McDonald, Best and Andersen to his home at Waiomatatini on the East Coast, so that records could be obtained of the traditional skills and technologies of his people, the Ngati Porou tribe (Dennis, 1996: 292). This was part of Ngata's explicit strategy to employ *nga rakau a te Pakeha* (the tools of Europeans) in the recording of old forms of knowledge and material arts – *nga taonga a o tipuna* or the treasures of ancestors – for use in the Young Maori Party's programme of economic and cultural invigoration. The team was joined by Ngata and Buck, and was assisted by many senior *kaumatua* (elders) in the district, who 'considered the recording of their arts and crafts and tribal lore as a matter of considerable importance' (New Zealand Film Archive, 1987). At towns with electricity, films from the Whanganui expedition were shown to the locals (New Zealand Film Archive, 1987). Buck had suggested this in a letter to McDonald, noting that viewing the films might 'perhaps make them vie with one another to produce good stuff to put on record',[9] a strategy that was evidently successful.

During the expeditions, the team visited towns and small, isolated rural settlements, recording elements of 'traditional' life performed especially for the camera along with more 'modern' daily activities including the sheep and dairy farming that were part of Ngata's land development initiatives (Craig, 1964: 188). The material obtained on these journeys, in the form of notes, recordings, photographs, film footage and artefacts, was to be employed both for the enrichment of anthropological knowledge, and in the Young Maori Party's efforts to ensure that the ways of their ancestors were brought forward into Te Ao Maori, the contemporary Maori world.

In 1923 Ngata succeeded in passing legislation to establish the Board of Maori Ethnological Research (later the Maori Purposes Fund Board), to which Best, Buck and Ngata himself were all appointed. The Board used monies from unclaimed Maori rentals and other sources to support the Polynesian Society (of which Best was now president) and its publications (Sorrenson, 1996: 361). It was decided that the publication of data obtained on the expeditions should be undertaken by the Polynesian Society, and £250 was donated toward costs by Ngata's Ngati Porou people (Craig, 1964: 202). The expedition *Bulletins* were published between 1924 and 1927, supported by Maori funding (Sissons, 1993: 40).

Ngata became Native Minister in 1928 and was able to launch the Young Maori Party's programme of land development and cultural regeneration on a grand scale. He had earlier considered writing a doctoral thesis on Maori social organization, using Rivers' genealogical method as a model (Sorrenson, 1982), but instead applied the results of his anthropological researches in office. A Maori Arts and Crafts Act was passed in 1926 and a School of Maori Arts and Crafts established at Rotorua. During Ngata's administration, the School became a centre for the revival of carving, weaving and other art forms recorded on the Dominion Museum expeditions. As Native Minister, Ngata also arranged financial assistance and supervision for Maori agricultural developments and the building of twenty-eight tribal meeting houses, many of which were carved by students of the Rotorua School (Sissons, 1998: 43). After resigning his ministerial post, Ngata remained in politics for a time, also publishing collections of Maori chants, articles on carving, and revising the Maori dictionary. His initiatives provided a focus for the revitalization and development of Maori arts and culture that continues today.

After the final East Coast expedition, in 1923, Buck met A. C. Haddon at a conference in Australia. Haddon expressed an interest in the team's work, particularly Andersen's studies of string games. In a letter to 'Tarawhai' (his pen name for Andersen), Buck asked his friend to send copies of his work to Cambridge.[10] Buck, who had recently been appointed to the Auckland Museum Council as inaugural chair of its Anthropology and Maori Race section, went on to become a research fellow at the Bernice P. Bishop Museum in Hawaii, where he wrote his first book *The Evolution of Maori Clothing*, published by the Polynesian Society in 1927. His classic text *The Coming of the Maori* (1949) was also based on meticulous, first-hand analyses of Maori material culture, reminiscent of the work of Haddon. For the rest of his career Buck carried out fieldwork as a professional anthropologist throughout Polynesia, focusing his research on Polynesian migrations and material culture. He considered his *maoritanga* ('Maori-ness') to be of central importance in his work, writing to Ngata that, 'In Polynesian research, it is right and fitting that the highest branch of the Polynesian race should be in the forefront and not leave the bulk of the investigation to workers who have not got the inside angle that we have.'[11] In 1932 he was awarded a visiting professorship at Yale University, and became the director of the Bishop Museum in 1936. Over the years Buck maintained a lengthy correspondence with Apirana Ngata, whom he visited in New Zealand shortly before his death at Honolulu in 1951.

Ralph Piddington and Maori Anthropology

The museum- and field-based anthropology pioneered by Buck, Ngata and the Dominion Museum team, in correspondence with Rivers, Haddon, Pitt-Rivers and other British and American scholars, did not continue when anthropology gained a foothold in New Zealand universities. Rather, it spiralled off into Ngata's initiatives for Maori cultural and economic invigoration, and Buck's Hawaii-based Pacific research.[12] One of Haddon's students from Cambridge, the New Zealander Henry Devenish Skinner, was appointed to Otago University in 1919 as Curator of the Museum and the country's first lecturer in ethnology but his interests lay more in the field of archaeology than contemporary ethnography, and he did not continue the ethnological tradition established by Buck and others. Another New Zealander, Raymond Firth, conducted his seminal study of Maori economics in the 1920s under the supervision of Malinowski, but his teaching career began in Australia and developed in England, and it was only later that his influence was felt in New Zealand anthropology.

It was not until the 1950s and 1960s, when a new generation was trained under Ralph Piddington – an Australian student both of Malinowski and Radcliffe-Brown – that Maori participation in anthropology again began to flourish. Piddington, who established the first university department of anthropology in the country at Auckland, introduced the teaching of Maori language as an academic subject and created an environment that welcomed and nurtured young Maori scholars, several of whom went on to complete doctorates at Oxbridge and in the US. Taking up lecturing positions in New Zealand, these Maori anthropologists helped initiate political developments that saw their people reassert with new vigour old claims to redress for injustices meted out during colonization.[13] Their work, which was concerned with advancing Maori interests at home perhaps more than it sought to engage in international debates within anthropology, has nonetheless attracted scholarly attention among those studying the discipline's role in New Zealand's post-colonial political situation.

Steven Webster (1998), an American anthropologist who worked in New Zealand from the 1970s, credits Piddington and his students with having crystallized a 'culturalist ideology', which, he argues, has long characterized Maori anthropology in New Zealand. Identifying a substantial gap between Maori culture as it is conceived and presented by scholars of Maori studies and anthropology (and indeed by many Maori in general), and 'culture' as the day-to-day lives of Maori people, Webster traces the roots of this 'ideological separation' back to Piddington's personal brand of socio-cultural anthropology, and beyond him to the Young Maori Party's programme for Maori cultural renewal. This laid the ground for Piddington's theory of culture, which in turn, according to Webster, led many of his students to similarly reify Maori culture in an essentialist and ahistorical fashion, and to ignore the political-economic forces that, he considers, are integral to the culture as it is actually lived. This idealized vision of Maori culture was then mobilized during the 1970s and 1980s in the period of political and cultural activity that Webster (with others) characterizes as the 'Maori Renaissance' (see also Walker, 1990). After outlining some key events in this period, which had a substantial impact on the discipline of anthropology as practised in New Zealand, I will consider what is elided in political economic explanations such as Webster's, namely the conceptual continuities evident both in the way in which Maori and others have rendered their 'culture', and in the assumptions which anthropologists have often deployed in dealing with these renditions.

Maori Anthropology from the 1970s

Throughout the twentieth century many Maori kin-groups protested the loss of their lands and economic base but from the late 1960s these demands gained impetus with the emergence of world-wide movements that sought to gain political equality for marginalized groups in liberal democratic polities. Debates raged in the academic and popular media, and the political landscape of New Zealand was significantly altered as Maori achieved a vocal presence on the national and international stage. Some of the most prominent and influential leaders in this period were Maori anthropologists, including Robert Mahuta, Pita Sharples, Hugh Kawharu, Pare Hopa, Hirini Mead, Pat Hohepa and Ranginui Walker, who, like Ngata before them, used their anthropological training to establish initiatives to ensure the continuation and revitalization of the Maori language and cultural traditions, and to theorize ways in which *maoritanga* could (and should) endure in a society dominated politically, economically, and numerically by the descendants of colonial settlers (see, for example, Kawharu, 1975; Walker, 1981, 1984). At the same time, activists like Titewhai Harawera challenged the right of non-Maori to commentate on Maori issues, reserving particular condemnation for foreign and *pakeha* scholars. Key events in this political mobilization included the Land March in 1975, when thousands of Maori and supporters walked the length of the country to protest the alienation of Maori land through colonization; the formation in that same year of the Waitangi Tribunal, a quasi-judicial government body investigating Maori claims for the loss of tribal resources and lands; violent protests against apartheid during the South African Springbok rugby tour in 1981; and *Te Maori*, the 'blockbuster' exhibition of Maori carvings from the collections of New Zealand museums, which toured the US from 1984. In anthropology, one effect of this period of vigorous activism was that ethnographic studies of present-day communities gave way by and large to practical initiatives, engagements with post-colonial theory, and excavations of historical material for evidence both of traditional ways of being and of past injustices. These developments also contributed to the departure of Maori from the discipline – many because they felt their intellectual resources better spent elsewhere[14] – and the redirection of the research interests of many *pakeha* anthropologists, for whom the political climate made contemporary ethnography 'too hot to handle'.

Developing alongside and in support of this activism, as Webster argues, was a particular image of Maori culture that has since become institutionalized in government policy and university departments, and which remains highly influential in present-day New Zealand. Webster has traced in detail the development of this version of *maoritanga* and its rise to the status of orthodoxy through the writing and activities of Piddington's students and others. Here, I wish to examine his analysis of the 'reification of [Maori] culture in idealist terms' in relation to other critiques of the way in which culture has been objectified, particularly those of Allen Hanson (1989) and Adam Kuper (2003).

Critiques of Cultural Reification

Hanson and Kuper's articles, alongside Webster's book, are among a number of analyses published since the late 1980s that have presented Maori (among other 'indigenous' cultures) as a discursive artefact, the hybrid offspring of more than one system of knowledge. Webster describes this 'popular stereotype' of *Maoritanga* as incorporating

> kin and community solidarity; respected and authoritative elders; public ceremonial and ritual symbolism in *hui* at *marae* (gatherings at kin-based meeting places including meeting-houses decorated with symbolic carvings, weavings, and paintings); generosity and sharing of resources; Maori language as mother tongue; harmony with the natural world; and profound spirituality centred in notions of *tapu* ('taboo'), *mana* ('prestige'), and *wairua* ('spirit'). (Webster, 1998: 29)

Webster is one of the few scholars to have written explicitly about the apparent gulf between this image of Maori culture and what he and others regard as the realities of contemporary Maori life. Indeed, New Zealand-based anthropologists have rarely reflected publicly on what to many observers seems a curious segregation. His account is persuasive, for one can indeed recognize two ways of studying things Maori in the literature of anthropology and related disciplines. The first is practised mainly by sociologists in New Zealand and by some anthropologists based in America or Europe who visit the country to conduct fieldwork. These scholars take 'Maori culture' to be the contemporary situation of Maori people as (culturally marked) members of a modern liberal democratic society. The second, and by far the most commonplace approach within New Zealand, identifies 'Maori culture' as a range of tenets, practices and 'spiritual' principles (much as Webster has listed above) said to have been handed down from ancestors and which, it is claimed, remain fundamentally relevant in present-day 'Maori society'. This second form of scholarship has dominated anthropology by and about Maori for much of its history in New Zealand, and is now the principal means by which Maori culture is defined by scholars in those university departments that have absorbed the exodus from the discipline, namely Maori Studies and Education, as well as in schools, government departments and other institutions charged with implementing cultural policy.[15] Although it has not been much examined by scholars in New Zealand, the apparent gap between these so-called 'real' and 'ideal' images of Maori culture has attracted considerable attention from politicians and others who disagree with government initiatives (including legislation) that single out Maori for 'special' or favourable treatment.[16]

In recent anthropological analyses conducted *outside* New Zealand, however, the 'reification' or 'objectification' of culture – the self-conscious reflection of a group of people on their differences from others – has often been interpreted as a sign of anthropological intervention. As I have argued elsewhere (Henare, 2007), analyses of cultural change (particularly those dealing with 'Fourth -World' societies or former colonies), have taken cultural reification as a practice with a distinctively Euro-American pedigree. In his article 'The Return of the Native', Adam Kuper launched a searing attack on what he describes as 'the indigenous-peoples movement' worldwide, noting how 'discredited old arguments may lurk behind new words' (2003: 389). Asserting that the origins of the discourse of indigeneity lie in European romanticism and early anthropological notions of race, culture and ethnicity, Kuper argues that:

> New identities are fabricated and spokespeople identified who are bound to be unrepresentative and may be effectively the creation of political parties... These spokespeople demand recognition for alternative ways of understanding the world, but ironically enough they do so in the idiom of Western culture theory.

What is entirely precluded by this assessment, of course, is the possibility that the concepts mobilized in such claims – in the very words 'culture', 'world view', and 'indigeneity' – might have different connotations among Maori from those familiar to metropolitan anthropologists. There is a problem of translation here, one that is exacerbated rather than helped by anthropologists sharing the language of their ethnographic subjects. When one works in a place where the language is different from one's own, it is easier to allow that the concepts one encounters in the field might be entirely unfamiliar. In New Zealand, where the first language of most Maori is English, and where Maori 'culture' is marked by long-standing engagements with anthropology, it is almost irresistible to assume that when people invoke terms such as 'the Maori world view', they are simply parroting old-fashioned anthropology. However, what if Maori continue to draw on a conceptual repertoire that not only includes notions drawn from anthropological discourse but also those that could be described as distinctively Maori?[17] After all, if romanticism and race theory have survived decades of sustained theoretical critique, why should concepts like 'Te Ao Maori' (typically rendered in English as 'the Maori world') not have similarly endured through the vicissitudes of post-colonial history?

Like Kuper, Allen Hanson negates this possibility, arguing that:

> Anthropologists and historians have become acutely aware in recent years that 'culture' and 'tradition' are anything but stable realities handed down intact from generation to generation. Tradition is now understood quite literally as an invention designed to serve contemporary purposes… (Hanson, 1989: 890)

Hanson argued that significant aspects of Maori cultural 'tradition' are in fact scholarly fabrications invented by nineteenth-century anthropologists, including Elsdon Best, S. Percy Smith and Peter Buck, and that an idealized version of Maori culture has been upheld more recently by scholars in departments of anthropology. These 'inventions', he asserts, are subscribed to by contemporary Maori for political and economic ends, specifically for the purpose of enhancing their own position in New Zealand society. Hanson debunks these mythologies with evidence drawn from archaeology and history. While claiming that his own intentions are apolitical in that 'the invention of culture is no extraordinary occurrence but an activity of the same sort as the ordinary, everyday process of social life' (Hanson, 1989: 899) he immediately qualifies this disclaimer with the following observation:

> [T]here must nevertheless be something distinctive about culture invention. It is, after all, much too strong a phrase to use for everyday social reproduction. As a first approximation, it might be said that inventions are sign-substitutions that depart some considerable distance from those upon which they are modelled, that are selective, and that systematically manifest the intention of some political or other agenda. (Hanson, 1989: 899)

This brings Hanson's argument close that of Kuper and Webster, both of whom regard the elevation of an idealized and reified image of indigenous/Maori culture as insidious. While he argues on one hand that 'the analytic task is not to strip away the invented portions of culture as inauthentic, but to understand the process by which they acquire authenticity', on the other hand Hanson provides, both explicitly and by example, the analytic tools with which to demolish such cultural 'inventions'.[18]

Te Ao Maori Versus the Maori World

Although the arguments sketched above sometimes gesture toward Roy Wagner's elaboration of *The Invention of Culture* (1981), they owe much more in theoretical terms to Hobsbawm and Ranger's *The Invention of Tradition* (1983). Certainly, the subtleties of Wagner's discussion have been lost in analyses that present the objectification of culture as a mark variously of modernization, 'cross-cultural hybridity', or the contamination of what were once unreflexively lived (and therefore organically authentic) lives. Whereas Wagner's insights were inspired by theorizations of innovation on the part of his Daribi informants, other accounts of cultural 'invention' appear to regard creativity as a distinctively Euro-American proclivity.[19] That such analyses work to undermine the basis for certain kinds of culture-based claims is clear (Henare, 2007) but my interest here is less in their political effects than what they betray about the persistence of certain assumptions underlying the anthropological endeavour.

Critics of the 'reification' of Maori culture have defined the notion of a different Maori 'world' – one that endures alongside or within the everyday world of modern New Zealand society, as an illusory (and possibly dangerous) discursive construction. Webster, for example, explicitly attacks what he calls 'the neo-Kantian model of "two worlds"' (1998: 126), arguing that the concept of multiple realities is little short of bizarre. 'Contrary to a popular preconception', he writes, 'Maori do not live in one or both of "two worlds" and probably never have; everyone lives in this world' (Webster, 1998: 26). Elsewhere he identifies such constructions as 'highly ideological, tending to obscure and mystify certain conflicts that need to be resolved in what is really just one world' (Webster, 1998: 118). Returning to Viveiros de Castro, Webster's claim may be identified with the mono-naturalist/multiculturalist position in anthropology, in which there is room for only one 'world', one universal ontology.

In a recent publication that seeks to divert anthropology from its predominantly epistemological orientations, the editors argue that such a position is no longer tenable in a discipline that seeks to take its informants seriously. In the persistent dualisms of object versus subject, material versus culture, world versus representation, they note, is embedded an unavoidable connotation – the implicit contrast between appearance and reality: 'In its traditional rendition (representation = appearance, world = reality)', the authors note,

> it is just this extra implication that makes the power of dualism so pernicious to anthropological thinking. For if cultures render different appearances of [a single] reality, it follows that one of them is special and better than all the others, namely the one that best *reflects* reality. And since science – the search for representations that reflect reality as transparently and faithfully as possible – happens to be a modern Western project, that special culture is, well, ours. (Henare, Holbraad and Wastell, 2007)

This attitude is evident in each of the critiques of cultural 'reification' outlined above. Only the reflexive scholar can cut through the mythologies to see reality. Rather than allowing that Maori might conceive things differently, the anthropologist dismisses their accounts as misguided, illusory, and, worst of all, indicative of the tragic irony that such people can only proclaim their difference using 'our' language, because they are no longer themselves.

Which brings me to one final observation. It is easy to argue that early Maori and Europeans held very different assumptions about the nature of reality but a perennial problem for anthropologists is how to account for what happened to such conceptions as relationships between these people developed. A tacit assumption in the discipline, which is becoming increasingly explicit both in anthropology and in broader New Zealand society, holds that since the arrival of Europeans Maori have progressively 'lost' the qualities that originally marked them out as different from the European settlers. Just as they have intermarried, it is assumed, their concepts have become hybridized and diluted. Questions are accordingly raised by politicians about the sense (and ethics) of segregating Maori interests from those of the general population and the idea that Maori should claim a separate identity at all is under vigorous attack. As a Marxian theorist, Webster for one would almost certainly not wish to be allied with such views, which are usually voiced by those on the right of New Zealand's political spectrum – Webster's targets are middle-class elites of every ilk and he champions the Maori working class against their 'patrons', Maori and *pakeha* alike. Yet his argument, like those of Hanson and Kuper, has similar implications.

The idea that peoples such as Maori are simply borrowing from the conceptual toolbox of anthropology (or of the West in general) for the sake (however laudable) of self-determination obscures another possibility – that they are using the language of social theory in an attempt to articulate (possibly untranslatable) concepts of their own. One reason why scholars might find it difficult to take their pronouncements seriously is not simply that the terminology deployed can appear somewhat dated – the often-used 'Maori world view' springs to mind – but that we are unwilling to acknowledge even the *potential* alterity of what is being put forward. There is an idea about that incorporation into global capitalism and modernity involves a net conceptual loss on the part of the less powerful, resulting in a tragic though inevitable decline into inauthenticity, a kind of conceptual miscegenation. In this sense the very act of incorporating people into the *anthropos* – the 'family of mankind' – long regarded as the philosophical baseline of anthropology, entails deploying assumptions built on specific understandings of relatedness. Notwithstanding that such ideas have been thoroughly unpacked in 'the new kinship studies', or that terms such as 'race' have been largely rejected (at least in the academy), a residual biologism remains evident at the heart of contemporary anthropology. There are very good reasons for insisting on the authority of this conception, connected historically to the abolition of slavery, the rights of women and other moral and political imperatives. It comes, however, with a degree of hubris about the power and efficacy of some conceptions over others, a sort of philosophical 'survival of the fittest'. The strength of such moral convictions undermines analytic flexibility, preventing us from taking seriously other peoples' accounts of reality.

Notes

1. Living in close contact with Maori from 1815, the missionary Thomas Kendall, for example, found himself embroiled in Maori cosmologies and left his family for a time to live with a local woman, the daughter of a *tohunga* or priestly expert. He was powerfully affected by his exposure to Maori ways of being, writing 'I have been so poisoned with the apparent sublimity of their ideas, that I have been almost completely turned from a Christian to a Heathen.'
2. In debunking the claim that Maori, along with other indigenes, are natural conservationists, for example, Eric Kolig writes that 'such notions have become part of the "political myths" constructed by [indigenous people] or assimilated from dominant western society. Convinced as they may be of the validity of this notion, it remains dubious whether there is any basis for this belief in historical (or pre-historical) fact' (Kolig, 2002: 109).
3. Rivers was returning to Europe, after attending the 1914 meeting in Australia of the British Association for the Advancement of Science and spending several months in the northern New Hebrides.
4. *Pakeha* describes New Zealanders of European descent. The term is rendered in lower case throughout this text to denote a personal preference for its use as an adjective, rather than as a proper noun designating a pseudo-ethnic category.
5. See also the video of the footage gathered on this trip (New Zealand Film Archive, *Te Hui Aroha ki Turanga/Gisborne Hui Aroha*, 35 mm, black and white, silent, 10 minutes).

6. New Zealand Film Archive, *He Pito Whakaatu i te Hui i Rotorua/Scenes at the Rotorua Hui*, 35 mm, black and white, silent, 24 minutes.
7. New Zealand Film Archive, *He Pito Whakaatu i te Noho a te Maori i te Awa o Whanganui/Scenes of Maori Life on the Whanganui River*, 35 mm, black and white, silent, 48 minutes.
8. Letter from McDonald (Acting Director, Dominion Museum) to the Under Secretary for Internal Affairs, 3 December 1923, Dominion Museum Archives 13/27/29.
9. Correspondence: Te Rangi Hiroa to James McDonald, 6 February 1923, MONZ Archives DM 11 3/3.
10. This request was evidently fulfilled, for Haddon's personal copy remains today in the library named after him in Cambridge. Letter from Te Rangi Hiroa (Peter Buck) to Tarawhai (Johannes Andersen), 2.10.1923, Alexander Turnbull Library, MS papers 48.
11. Buck to Ngata, from Auckland, 8 March 1927 (in Sorrenson,1986: 48; see also Condliffe, 1971: 151).
12. Buck did however continue to publish on Maori topics, including a jointly-written article on Maori kinship which appeared in *American Anthropologist* in 1940 (Aginzky and Te Rangihiroa, 1940).
13. For a detailed discussion of this period, see Orange, *The Treaty of Waitangi*, 1987, pp. 243–54.
14. This observation is gleaned from a series of interviews I conducted with Maori who trained as anthropologists from the 1950s, including the pioneer of Maori language education Dr Pita Sharples, now an MP and leader of the new Maori Party.
15. One of innumerable examples that could be presented here is drawn from a Department of Justice report, which notes: 'Maori do not and never have accepted the system of a closed world. They believe the spiritual realm interacts with the physical world and vice versa... To understand this concept, it is important to look at how Maori see the world that they live in. James Irwin describes the 'Maori World View' as a three-tiered structure...'
16. In a speech to the Orewa Rotary Club in 2004, Don Brash, Leader of the Opposition, asserted: 'It is bizarre that, in a society where the Prime Minister refuses to allow grace to be said at a state banquet ... we fly Maori elders around the world to lift tapu and expel evil spirits from New Zealand embassies; we allow courts to become entangled in hearings about the risks to taniwha of a new road or building; we refuse to undertake potentially life-saving earthworks on Mount Ruapehu lest we interfere with the spirit of the mountain... We are becoming a society that allows people to invent or rediscover beliefs for pecuniary gain.'
17. Viveiros de Castro has examined this issue in relation to concepts of the body, in a paper presented to the 2002 annual meeting of the American Anthropological Association.
18. Although he later acceded that the term *invention*, 'when applied to culture and tradition is a systematically misleading expression that should not be

perpetuated', Hanson has continued to defend 'the overall theoretical approach, and the particular analysis of Maori culture, contained in the essay' (Hanson, 1991: 450; and see also Hanson, 1997).

19. See Henare, Holbraad and Wastell (2007) for a more detailed discussion of the re-invention of Wagner's argument by theorists of the invention of tradition.

Bibliography

Aginsky, B. and Te Rangi Hiroa (Peter Buck) (1940), 'Interacting Forces in the Maori Family', *American Anthropologist*, 42: 195–210.

Anon. (1919), *Annual Report of the Dominion Museum,* Wellington: Dominion Museum.

—— (1921), *Annual Report of the Dominion Museum*, Wellington: Dominion Museum.

Buck, P. (1949), *The Coming of the Maori*, Wellington: Whitcombe & Tombs.

Condliffe, B. (1971), *Te Rangi Hiroa: the Life of Sir Peter Buck,* Christchurch: Whitcombe & Tombs.

Craig, E. W. G. (1964), *Man of the Mist: A Biography of Elsdon Best,* Auckland: A. H. & A. W. Reed.

Curnow, J. (2005), 'Te Rangikaheke, Wiremu Maihi – 1896'. *Dictionary of New Zealand Biography*, updated 7 July 2005. Available online at http://www.dnzb.govt.nz/.

Dennis, J. (1996), 'McDonald, James Ingram', in Department of Internal Affairs (New Zealand), *Dictionary of New Zealand Biography*, Vol. 3, Auckland: Auckland University Press.

Gibbons, P. J. (1992), *'Going Native': A Case Study of Cultural Appropriation in a Settler Society with Particular Reference to the Activities of Johannes Andersen in New Zealand During the First Half of the Twentieth Century.* DPhil thesis, University of Waikato.

Grey, G. (1885), *Polynesian Mythology and Ancient Traditional History of the New Zealand Race as furnished by their Chiefs and Priests*, Auckland: H. Brett.

Hanson, A. (1989), 'The Making of the Maori: Culture Invention and its Logic', *American Anthropologist*, 91: 890–90.

—— (1991), 'Reply to Langdon, Levine, and Linnekin', *American Anthropologist*, 93(2): 449–50.

—— (1997), 'Empirical Anthropology, Postmodernism, and the Invention of Tradition', in M. Mauzé (ed.), *Present is Past: Some Uses of Tradition in Native Societies*, Lanham: University Press of America.

Henare, A., (2007), 'Taonga Maori: Encompassing Rights and Property in New Zealand', in A. Henare, M. Holbraad and S. Wastell (eds), *Thinking Through Things*, London: Routledge.

—— (2005), *Museums, Anthropology and Imperial Exchange*, Cambridge: Cambridge University Press.

Henare, A., Holbraad, M. and Wastell, S. (2007), 'Thinking Through Things', in A. Henare, M. Holbraad and S. Wastell (eds), *Thinking Through Things*, London: Routledge.

Hobsbawm, E. and Ranger, T. (eds) (1992 [1983]), *The Invention of Tradition*, Cambridge: Cambridge University Press.

Kawharu, H. (ed). (1975), *Conflict and Compromise: Essays on the Maori since Colonisation*, Wellington: Reed.

Kolig, E. (2002), 'Guardians of Nature or Ecologists of the Stomach? The Indigenous Cultural Revival in New Zealand, Resource Use and Nature Conservation', in E. Kolig and H. Müchler (eds), *Politics of Indigeneity in the Pacific: Recent Problems of Identity in Oceania*, Münster: Lit Verlag.

Kuklick, H. (1993), *The Savage Within: the Social History of British Anthropology, 1885–1945*, Cambridge: Cambridge University Press.

Kuper, A. (2003), 'The Return of the Native', *Current Anthropology*, 44(3): 389–95.

—— (1996), *Anthropology and Anthropologists: the Modern British School*, London: Taylor & Francis.

Luomala, K. (1952), *Necrology*, Honolulu: Bishop Museum Press.

New Zealand Film Archive (1987), *Films by James McDonald* (screening programme). Wellington: The New Zealand Film Archive.

Sahlins, M. (1999), 'Two or Three Things that I Know about Culture', *Journal of the Royal Anthropological Institute*, 5: 399–421.

Sissons, J. (1998), 'The Traditionalisation of the Maori Meeting House', *Oceania*, 69(1): 36–46.

—— (1993), 'The Systematisation of Tradition: Maori Culture as a Strategic Resource', *Oceania*, 64(2): 97–116.

Sorrenson, M. P. K. (1996), 'Apirana Turupa Ngata', *Dictionary of New Zealand Biography*, Vol. 3, Auckland: Auckland University Press.

—— (1992), *Manifest Duty: The Polynesian Society over 100 years*, Auckland: Polynesian Society.

—— (1986), *Na To Hoa Aroha From Your Dear Friend: The Correspondence between Sir Apirana Ngata and Sir Peter Buck 1925–50*, 3 vols, Auckland: Auckland University Press.

—— (1982), 'Polynesian Corpuscles and Pacific Anthropology: The Home-Made Anthropology of Sir Apirana Ngata and Sir Peter Buck', *Journal of the Polynesian Society*, 91: 7–27.

Stocking, G. (1992), *The Ethnographer's Magic and Other Essays in the History of Anthropology*, Madison: University of Wisconsin Press.

—— (1987), *Victorian Anthropology*, New York: Free Press.

Strathern, M. (1990), 'Artefacts of History: Events and the Interpretation of Images', in J. Siikala (ed.), *Culture and History in the Pacific*, Helsinki: Suomen Antropologinen Seura (The Finnish Anthropological Society) Transactions No. 27, pp. 25–44.

Viveiros de Castro, E. (2002), 'O nativo relativo', *Mana*, 8(1): 113–48.

—— (1998), 'Cosmological deixis and Amerindian perspectivism', *Journal of the Royal Anthropological Institute*, 4(3): 469–488.

Wagner, R. (1981), *The Invention of Culture*, Chicago: University of Chicago Press.

Walker, R. (1990), *Kawhawhai tonu matou: Struggle Without End*, Auckland: Penguin.

—— (1984), 'The genesis of Maori activism', *Journal of the Polynesian Society*, 93(3): 267–81.

—— (1981), *Perceptions and Attitudes of the New Generation of Maoris to Pakeha Domination*, Winter lecture 5 August, University of Waikato.

Webster, S. (1998), *Patrons of Maori Culture: Power, Theory and Ideology in the Maori Renaissance*, Dunedin: University of Otago Press.

–6–

The Second Nuclear Age[1]

Hugh Gusterson

> Meanwhile, the countries of the North harden their borders and stockpile weapons of mass destruction. After all they have to make sure it's only money, goods, patents and services that are globalized.
>
> Arundhati Roy

Anthropology largely missed the First Nuclear Age. Consequently there has been a big hole where there could have been a conversation between anthropologists and political scientists about international relations and the Cold War.[2] In the early years of the Cold War European and American anthropologists were still predominantly focused on the description of the social and cultural dynamics of small-scale societies as they emerged from the colonial era. Any American anthropologists with an interest in larger political questions were, as Laura Nader (1997) and David Price (2004) have argued, given second thoughts in an atmosphere of McCarthyism that largely depoliticized the humanities and social sciences in the US for a generation.

From the late 1960s anthropologists did broach larger questions of international political economy but their interest was generally focused on the social dynamics of colonialism and decolonization, transnational migration, industrialization and, most recently, globalization. Those engaged in repatriated anthropology have tended to concentrate on issues of social stratification, gender politics, and cultural pluralism rather than militarism and science.[3] Insofar as anthropologists have taken an interest in war in recent years it has been largely to debate whether war is a cultural universal and to write about some of the regional and civil wars generated by colonialism and decolonization. Given anthropologists' traditionally favoured research technique of participant observation in localized communities, the overarching global conflict, the Cold War itself, scarcely came into ethnographic focus; and, in the days before the anthropology of science, the laboratories and expert communities at the heart of Cold War culture lay outside the analytic territory of anthropology departments.

Thus in the 45 years of the First Nuclear Age, despite the huge social movements against the arms race in the 1950s and 1980s, we have only one major anthropological study of an anti-nuclear movement – Louise Krasniewicz's *Nuclear Summer* – and only one major study of a Cold War military-industrial facility: my own *Nuclear Rites*. Although the Cold War brought about a substantial US military presence in many countries around the world, the ethnographic literature on Turkey and the Philippines, for example, barely mentions the huge US military bases in those countries. Meanwhile anthropological writing on American culture during the Cold War scarcely references the fact that the US was engaged in a project as remarkable in its own way as the Egyptians' building of the Pyramids: the stockpiling, at a cumulative expense of $5.5 trillion,[4] rationalized by ideologies as bizarre as any we have ever discovered in rainforests and jungles, of enough nuclear weapons to end all life on earth many times over. Nor, with the exception of David Price's (2004) work, have histories of anthropology focused on the Cold War as an environment that shaped the practice and teaching of anthropology.

Many of the factors that impeded anthropologists from attending to nuclear weapons issues in the past now seem to be less salient: recent years have seen the rise of an anthropology of the state and of globalism, of the anthropology of science, of repatriated anthropology and of a reinvigorated anthropology of elites, as well as a new openness to multi-sited fieldwork and other deviations from our traditionally canonical method of small-scale participant observation. Anthropologists today are increasingly eager to understand global neo-liberalism, state bureaucracies, intra-state and inter-state wars, and ideologies of nationalism and modernity. This ensemble of interests, amounting to an emergent anthropological portrait of global society itself, will be incomplete without some attention to nuclear weapons and their place in human society. These weapons are important not only as material artefacts that might yet destroy global civilization as we know it but also as symbolic tokens the deployment of which regulates status hierarchies in the international system. Since the end of the Cold War there have been shifts in the international distribution of these weapons, in the treaty system that weakly regulates the rights and responsibilities of their owners, in the practices of their progenitors in the scientific community, and in our collective imagination of the threats the weapons embody and deter. These shifts, taken as a whole, define the transition from the first to the Second Nuclear Ages.

In this chapter I want to explore some of the properties of this emergent Second Nuclear Age and of the international system it defines, by way of a discussion of my fieldwork among American nuclear weapons scientists. I have been studying these scientists since 1987, when, in the waning light of the First Nuclear Age, I embarked on an ethnographic study of the Lawrence Livermore National Laboratory, the California facility whose scientists designed the nuclear warheads for the Polaris, MX, and Ground-Launched Cruise missiles. Since the end of the Cold War I have continued fieldwork amongst the scientists of the

Livermore Laboratory, but have expanded my research to other sites as well: the other American nuclear weapons laboratory at Los Alamos, where the bombs dropped on Hiroshima and Nagasaki were designed; the defence bureaucracy (encompassing Congressional aides, White House officials, and staffers in the Department of Energy and the Department of Defense) in Washington DC; and nuclear weapons networks in Russia.[5] This research aims to understand how American and Russian nuclear weapons scientists are adapting to the end of the Cold War and what sort of world order they and their counterparts elsewhere are creating in place of that conflict.[6]

Such research requires methodological innovation. For example, while my first fieldwork, for an ethnography of the Livermore Laboratory, did at least focus on a single localized site in the manner of classic ethnographies, it was a top secret site that was doubly inaccessible to me as a person lacking a security clearance and as a British citizen. In order to investigate the culture of the laboratory I was almost never allowed to visit, I developed a heterogeneous toolkit of approaches that I refer to as 'polymorphous engagement' (Gusterson, 1997). These ranged from extensive reading of archival and journalistic sources to several dozen life-history interviews in weapons scientists' living rooms and participant observation in settings one step removed from the laboratory itself: housing I shared with laboratory employees, Sunday church services in Livermore, offsite laboratory sports teams and so on.

My second fieldwork has decentred traditional fieldwork practices still further because I am investigating processes of negotiation that encompass the Congress and several different offices of the executive branch in Washington DC, one weapons laboratory in New Mexico and another in California, weapons scientists in Russia, and the diplomatic corps of several nations. George Marcus's (1995) celebrated notion of 'multi-sited fieldwork', intended to facilitate ethnographic study of global processes, is only partially adequate to charter such research as it is still too tethered to concrete spatial locales rather than to radically decentred processes of decision-making. To really understand nuclearism, however, we need an anthropology that can trace connections and interfaces between spatially dispersed communities of scientists, politicians, military officers, defence intellectuals, diplomats, bureaucrats and journalists. These networked connections will by no means be contained within nation-states.

The Second Nuclear Age

At this point it might be useful to introduce as a foil for my argument – and an illustration of the ideological stakes of our discussion – a recent article titled 'The Structure of the Second Nuclear Age' by the conservative Yale political scientist Paul Bracken.[7] Bracken argues that the First Nuclear Age, dominated by the rivalry of the two superpowers, was characterized by a strategic stability that he explains both in terms of game theory and the characteristics of the two superpowers themselves. He attributes the alleged stability of deterrence during the Cold War to the geometric predictability of two-player rivalries, and to the 'icy rationality' and 'noble internationalisms' (Bracken, 2003: 407) shared by American and Russian leaders linked as they were -- or at least as he says they were -- by an Enlightenment tradition of civility and reason. Bracken expects the Second Nuclear Age to be volatile and dangerous since the number of nuclear players will increase, rendering the nuclear game less stable and predictable, and because new and immature players in the Third World, using nuclear weapons as state-building symbols rather than as rational tools of security, will be driven by what he calls 'the kinds of ludicrous behavior that arise out of nationalism' (Bracken, 2003: 408). In other words for Bracken the hallmark of the Second Nuclear Age is the dangerous proliferation of nuclear weapons to what he calls 'small and economically unserious countries' seeking to change the international order. Writing before the 2003 invasion of Iraq by the US and the UK, he gives as an example of such a country 'Iraq, with a GDP 15 percent the annual revenues of Wal-Mart' (Bracken, 2003: 412). Bracken advocates missile defence and preemptive attacks on countries seeking nuclear weapons as ways of increasing the security of the West in the Second Nuclear Age.

In the account that follows I want to work my way toward a different characterization of the Second Nuclear Age. This account will also foreground nuclear proliferation as a defining theme of the new nuclear order, but will cast its analysis in terms of a critique of the American imperial project rather than uncritically adopting a neo-colonial vocabulary that ranks nations in terms of their evolutionary fitness to possess nuclear weapons. Rather than taking it for granted that international security is threatened by the rise of what policymakers and pundits have taken to calling 'rogue states',[8] I want to ask how the long-standing Western anxiety about nuclear proliferation has been recoded – both in discourse and in strategic doctrine – since the end of the Cold War and to enquire into not only the shifting structure of nuclear deterrence but also the relationship between nuclear weapons and the global order itself since the fall of the Berlin Wall. I also want to juxtapose the increasing anxiety in the West about non-Western nuclear weapons with the rise of simulations in the practice of contemporary nuclear weapons science, since the emergence of powerful new simulation technologies among the established nuclear powers in the 1990s enabled the negotiation of a new treaty regime *vis-à-vis* nuclear weapons and offered American and other Western policymakers a symbolically rich new way to imagine the military hierarchy of nations.

Nuclear Weapons Science after the Cold War

The US conducted its last nuclear test on 23 September 1992. In the 1980s the US government had ignored pressure for a nuclear test ban treaty from the international anti-nuclear movement and the Soviet government. By 1992, however, the Cold War won, the US had no plans for new nuclear weapons, and American war planners were more worried about the weapons of a collapsing Soviet Union falling into the wrong hands than about falling behind in an arms race. Meanwhile, freed by the end of the Cold War, the non-nuclear countries of the world were becoming increasingly vocal about the nuclear powers' obligation under Article VI of the 1968 NonProliferation Treaty. In Article VI the nuclear powers had committed themselves 'to pursue negotiations in good faith on effective measures relating to cessation of the nuclear arms race at an early date and to nuclear disarmament.'

In this context Congress passed the Hatfield-Exon-Mitchell Amendment to the budget in 1992. The amendment mandated a nine-month pause in nuclear testing to be followed by negotiations for a permanent test ban treaty. As many as eighteen nuclear tests might be allowed after the testing pause but they were to be preparatory to a nuclear test ban, not a continuation of nuclear business as usual. The amendment also gave the President discretion to resume nuclear testing indefinitely if any other country tested a nuclear weapon. George H. W. Bush signed the amendment into law partly because he was assured by his advisors that the Chinese would conduct a nuclear test.

The Chinese did indeed do another nuclear test but, by then, Bill Clinton was President and he decided to forego further nuclear testing. The thinking in the Clinton White House was that, with a stockpile of 25,000 nuclear weapons, the US did not need to develop new nuclear weapons; but it did need to strengthen the international regime preventing the proliferation of nuclear weapons to other countries because the spread of nuclear weapons technology now nearly 50 years old could render smaller countries immune to US military dominance. (As the Indian General Sundarji famously said, the lesson of the 1991 Gulf War was not to risk war with the US unless you had nuclear weapons.) The single most effective non-proliferation measure, the Clinton Administration decided, was a comprehensive test ban treaty, because it would be difficult for new countries confidently to develop nuclear weapons if they could not test them, while the US, with over 1,000 nuclear tests under its belt, had the most thoroughly tested nuclear stockpile in the world.

The British government was very unhappy about this because Britain was dependent on the American nuclear test site for its own nuclear testing programme. When President Bush had signed into law the testing moratorium in 1992, the next test to take place at the Nevada Nuclear Test Site was Icecap, a British test on which most of the allotted $70 million had already been spent. This was to have been the twenty-fifth British nuclear test in Nevada.[9] Today, tourists can clamber over the 157-foot tall tower built in preparation for the aborted test. The decaying tower has been preserved as a sort of monumental ruin of the First Nuclear Age. When Bill Clinton came to power, the British pressed hard for a resumption of nuclear testing, with Secretary of State Warren Christopher making their case for them in the Clinton Cabinet. Christopher was overruled. In the complex negotiations that dragged on for another 3 years until the exact language of the nuclear test ban treaty was finalized, the British aligned themselves in the internecine struggles within the Clinton Administration with those hard-line elements of the Pentagon and the nuclear weapons complex that were pushing to allow very low-yield nuclear tests or occasional tests to validate the performance of ageing warheads in the stockpile. The British lost these diplomatic struggles but they did succeed – to the rage of American arms controllers – in inserting into the final treaty a poison pill clause that said the treaty would not fully enter into force unless signed and ratified by every single one of forty-four countries deemed capable of building nuclear weapons (Johnson, 1996).

It would be hard to exaggerate the devastating effect the end of nuclear testing had on morale at the American nuclear weapons labs. In the words of George Miller, Director of Weapons Development at the Lawrence Livermore National Laboratory, 'the organization we evolved to efficiently design, produce, test, and field new weapons, along with our entire rewards system, was all swept away by the end of the Cold War' (Miller, quoted in Kitfield, 1998: 2). In a very real sense nuclear tests were the main product of the nuclear weapons labs during the First Nuclear Age. The staging of these tests provided the main source of revenue for the labs, integrated different organizational teams in a shared project, and regulated status hierarchies while generating cascades of new knowledge. Without tests, many weapons designers and lab managers wondered what the labs would do now and how they would be able to guarantee the continued reliability of ageing weapons they could no longer test. In this context many scientists elected to take early retirement and, in the decade after the end of the Cold War, the number of core designers at Livermore and Los Alamos fell by about 50 per cent. The morale of those who stayed took a nosedive in a way that is nicely evoked in a song performed for me over a round of beer by a group of weapons scientists at Livermore. It is sung to the tune of the folk song *Those Were the Days*:

Sitting in my office after hours
Contemplating many yesterdays
I recall when A-bombs bloomed like flowers
Over Frenchman's Flats one sunny May

Those were the days, my friend
We thought they'd never end
We'd shake the earth and turn the night to day
We had no Cold War blues
We knew we couldn't lose
Those were the days, oh yes, those were the days.[10]

In the early 1990s, as the Clinton Administration was negotiating internationally for a nuclear test ban treaty, it was conducting parallel and intertwined negotiations with its own nuclear complex over the terms of the weapons labs' acquiescence in a permanent end to testing. The labs' main leverage here consisted in the fact that their directors would have to testify to the Senate in favour of such a treaty if it were to stand any chance of being ratified. At a key May 1993 meeting held inside a SCIF (Secure Compartmented Information Facility) in Washington DC, the lab directors struck a bargain with Hazel O'Leary, Clinton's Secretary of Energy, agreeing to co-operate with the test ban so long as they got as much money to simulate nuclear testing as they had received before to develop and test new weapons. Thus was born the Science Based Stockpile Stewardship programme, currently funded at $6.5 billion per year, which enabled the intellectual and organizational revitalization of the weapons labs in the 1990s.

The key components of the stockpile stewardship programme are as follows:

- *Dual Axis Radiographic Hydrotest Facility* (DARHT) – a Los Alamos accelerator facility that takes X-ray snapshots of imploding weapons in which the uranium or plutonium has been replaced with a simulant. Its tests show whether the uranium or plutonium would have been compressed symmetrically by the conventional explosives in a bomb as would be required if the chain reaction were not to 'fizzle'.[11]
- *National Ignition Facility* (NIF) – a $4 billion laser fusion facility, still under construction, whose 192 separate laser beams will combine to enable scientists to create within the lab temperature and pressure regimes such as those found within stars and hydrogen bombs. This is partly funded by the British taxpayer.[12] It is designed to enable scientists to understand and predict better the conditions in the final stages of a thermonuclear explosion and to investigate the effects these environments of intense heat, pressure and radiation flux have on specific materials. The French are, in consultation with their American counterparts, building their own version of NIF at a cost of over 2 billion Euros. This is the Megajoule laser, near Bordeaux.[13]
- *Subcritical tests* – underground tests conducted at the Nevada Nuclear Test Site in which small coin-sized pieces of plutonium are shocked with conventional explosives but do not go critical. Individual tests can cost as much as

$25 million, and the US has been conducting about three tests each year. These subcritical tests enable scientists to refine equations of state for plutonium and to explore changes in the behaviour of plutonium as it ages. The British have also participated in this programme.[14]

- *The Accelerated Strategic Computing Initiative* (ASCI) – a programme spanning all three nuclear weapons labs and five major universities, to develop a new generation of computer hardware and software to simulate the performance of nuclear weapons and analogous phenomena such as supernovae. The new generation of massively parallel computers, which includes what is currently the most powerful computer in the world (BlueGene/L) at Livermore, will replace old two-dimensional codes with new three-dimensional codes that integrate experimental data from the other components of the stockpile stewardship programme with the results of decades of nuclear testing.[15]

The rationale for the Science Based Stockpile Stewardship programme is threefold. First, it is hoped that data from the experimental facilities about processes within nuclear explosions and about the behaviour of materials used in nuclear weapons will enable scientists to improve the codes they use to simulate and predict the performance of these weapons. Second, if questions arise about the continuing reliability of a component or material that is ageing, the new experimental facilities will enable scientists to judge how it behaves in conditions similar to those found in a nuclear explosion. Third, the new facilities provide a means, in the absence of nuclear testing, for recruiting, training and rehearsing a new generation of weapons scientists. One Livermore scientist, likening stockpile stewardship to a physics gym, said

> A good analogy might be, supposing I'm a professional tennis player: if I sit on my butt and don't do anything, pretty soon I'm not going to win any games... But what if I go to a gym and bat balls against a backboard? I'm saying there's degrees to which you can keep a person up to speed... Facilities like the [National Ignition Facility] are exercise machines for designers.

The weapons labs have long believed that the ultimate guarantee of the reliability of the US nuclear stockpile resides in the expert judgement of the weapons designers themselves. The Stockpile Stewardship programme is a way of providing the kinds of expensive, state-of-the-art facilities that will enable the weapons labs to retain experienced designers from the Cold War, recruit new ones to replace them as they age and retire, and keep building the judgement and intuition of these scientists.

A New Geopolitics

This lurch toward the virtual took place during a decade that saw the US and North Korea come to the brink of war in 1994 over North Korea's covert nuclear weapons programme,[16] India[17] and Pakistan join the nuclear club with their salvos of nuclear tests in 1998, and growing anxiety in the West about 'loose nukes' leaking from the former Soviet Union to state or sub-state actors hostile to the West. Thus, as the weapons labs were building their new stockpile stewardship facilities, they were also reformulating their geopolitical orientation, building collaborative programmes with the Russian scientists who had once been their sworn enemies and developing new well funded programmes of surveillance to interdict the acquisition of nuclear weapons by Third World nations or terrorist groups. By the early 1990s, the beginning of the Second Nuclear Age, American security planners began to fear the conjuncture of Russia's defeat in the Cold War and a hunger for nuclear weapons on the part of some developing countries, the so-called 'rogue states', and sub-national groups such as what we now know as Al-Qaeda. If the central danger of the Cold War order had been that of massive nuclear war between the superpowers, the central danger of the new world order, they believed, was the slow dribbling of Russia's nuclear assets to other countries and the proliferation of weapons of mass destruction in black and brown hands around the world.

Regarding the Russian situation, American nuclear analysts were horrified to learn that, in post-Soviet Russia, nuclear materials were rarely inventoried and were often, in the words of Frank von Hippel, a Princeton physicist who worked in the first Clinton White House, 'secured only by a padlock and a nervous young man'.[18] And sometimes the young man, receiving only intermittent material support from Moscow, would go off to forage for food and fuel. United States defence analysts were worried because they believed that enough weapons grade material remained in the former Soviet Union to build at least 100,000 more nuclear weapons. Meanwhile, by the early 1990s, some of Russia's nuclear weapons scientists were three months behind in their paycheques and were reduced to growing their own vegetables to feed themselves.[19] In June 1992 scientists at Arzamas-16, where the first Soviet atomic bomb was designed, even wrote to Boris Yeltsin threatening to go on strike if they were not paid promptly (Horrigan, 1992: 53–4). These were the people who, in the words of Senator Sam Nunn, were 'capable of demanding and receiving hundreds of thousands, if not millions, of dollars for services rendered to Third World countries or terrorist groups' (quoted in Taubes, 1995: 488). In a decade that saw the Oklahoma City bombing, the car bombing of the World Trade Center, a war against Iraq to prevent it from acquiring nuclear weapons and a tense diplomatic crisis with North Korea over its covert nuclear weapons programme, the weapons labs and defence bureaucrats in Washington DC were increasingly preoccupied with such concerns.

The American nuclear complex has sought to stabilize the Russian situation by partly turning the Russian nuclear complex into a client. By the late 1990s the American taxpayer was spending $656 million a year on the Materials Protection, Control and Accounting Project, which sends American experts to Russian nuclear facilities to install alarms and other security systems that prevent workers or thieves from removing nuclear materials and selling them on the black market. The US, together with the European Union, has also been funding the International Science and Technology Center outside Moscow, which funnels grants to Russian weapons scientists to do non-weapons research. Finally, in an effort both to keep Russian weapons scientists busy and to transfer some of their knowledge to their American counterparts, the US funded a series of joint experiments between American and Russian nuclear scientists, particularly on pulsed power. By 1998, when these exchanges were at their height, there were so many collaborative projects between the American and Russian weapons labs that a number of Los Alamos scientists had begun learning Russian. Russians became the most common foreign visitors to Los Alamos, finally surpassing in number the British weapons scientists who have for decades collaborated with American scientists in developing nuclear weapons. In Livermore, by 1997, out of 600 visits to the lab by 'sensitive foreign nationals' 500 were by Russians. Livermore scientists were working with forty-two institutes in Russia and making 300 trips to Russia a year. One senior Livermore scientist told me he estimated that, by 1999, 30 per cent to 35 per cent of Chelyabinsk-70's budget was being provided by the US.

At the same time, in the mid-1990s, Livermore and Los Alamos created large new directorates focused on counter-proliferation – directorates that now account for more than a quarter of the laboratories' budgets. As these directorates looked out from behind their barbed wire fences, the problem as they saw it was that it would not be possible to contain 'rogue states' and nuclear-armed terrorist groups through nuclear deterrence as it had been practised with the Soviet Union because terrorist groups did not have national territory against which nuclear retaliation could be threatened, whereas the leaders of 'rogue states' were, so it was said, unlike Soviet leaders such as Stalin, indifferent to threats to exterminate millions of their citizens.

The laboratories have responded to this perceived conundrum with two strategies: enhanced detection and an attempt to reformulate deterrence itself. As far as detection goes, the weapons laboratories have, for example, been developing new technologies that can sniff the presence of concealed nuclear weapons materials for use by border guards, by UN inspectors in Iraq and elsewhere, and by emergency teams searching for a concealed terrorist bomb in a US city. They are also working on air-dropped miniature sensors, which, covertly placed in another country, would detect vibrations or magnetic disturbances from weapons construction plants; they are developing lasers that can identify the chemical composition of emissions from suspicious facilities and they are developing new satellite technologies to measure heat or survey radioactive emissions from concealed nuclear facilities.

Some weapons scientists and defence intellectuals have also suggested that, in order to deter 'rogue states', the US should deploy new kinds of nuclear weapons and formulate new strategic doctrines. These proposed innovations have been driven by a concern that leaders of 'rogue states' would not care if they lost entire cities in the pursuit of their objectives and are, hence, not deterrable in the way the Soviets were, whereas public opinion in the US would not support use of a nuclear weapon in such a situation if it would cause high civilian casualties. As Bob Kelley, a Los Alamos scientist, put it, while speculating about a nuclear attack on Iraq: 'We would have to take tens of kilotons to the problem. If it's anywhere near Baghdad it would destroy the entire city. That's not a usable deterrent' (quoted in Shroyer, 1997: 38).

As one of Kelley's colleagues told me, 'there are some evil people in the world. They don't care about the survival of their own people. Those people need something else.' The 'something else' proposed by some at the weapons laboratories and in Washington is the 'Robust Nuclear Earth-Penetrator' a nuclear weapon designed to burrow underground before exploding to target the leaders of rogue states in their underground bunkers or to pre-emptively destroy 'weapons of mass destruction' cached in underground bunkers.[20] The assumption here is that, even if they do not care about anyone else's life, these leaders do want to save their own. This represents an interesting inversion of Cold War strategic doctrine: during the Cold War it was thought unwise to target the Soviet leadership in a nuclear war because their presence would be required to negotiate an end to the war. Now it is precisely the leadership that is to be targeted.

The Administration of George W. Bush has pushed Congress hard to fund development of the Robust Nuclear Earth-Penetrator and, in its Nuclear Posture Review of 2001 and its 2005 draft Doctrine for Joint Nuclear Operations, it reformulated American nuclear doctrine to make use of such weapons more plausible. These two documents provided for pre-emptive attacks on non-nuclear states if they were thought to be developing 'weapons of mass destruction', sought to de-emphasize the traditional firebreak between nuclear and conventional weapons, and spoke specifically of the imperative for the US to use any suitable weapon against underground caches of 'weapons of mass destruction.'[21] At the time of writing, Congressional funding for the Robust Nuclear Earth-Penetrator has been blocked for the second year in a row by an alliance of Congressmen led by David Hobson, a Republican deficit hawk from Ohio, and by Ellen Tauscher, the Democrat whose district includes the Lawrence Livermore National Laboratory. For the moment, if the Pentagon wants earth-penetrating weapons, they will have to rely either on conventional weapons or on the modest earth-penetrating capability of the existing B61-11 nuclear gravity bomb.

Conclusion: The Second Nuclear Age

Table 6.1 summarizes in schematic form the contrasts between the First and Second Nuclear Ages. In the First Nuclear Age, the primary axis of conflict was between the two superpowers, with their client states ranged behind them. Now the rhetoric of enmity between the superpowers during the Cold War has given way to one of uncomfortable and inconsistent amity and American defence planners are increasingly preoccupied with threats from what George W. Bush referred to as the 'axis of evil': Iran, Iraq and North Korea. Insofar as they worry about nuclear war, they worry about an attack on the US by a rogue state or terrorist group, or about a regional nuclear war in South Asia or the Middle East. In other words, to put it in the terminology of Wallerstein, the dominant scenarios for nuclear trouble in the Pentagon today are core-periphery or within the periphery, rather than between the superpowers of the old global core. In parallel with this structural shift, the US has (with the possible exception of the Robust Earth-Penetrator, which has yet to be funded) largely given up the frenetic development of new nuclear weapons that characterized the First Nuclear Age, and is instead focused on what Joseph Masco (2004) calls 'nuclear gerontology' – the maintenance of an ageing stockpile inherited from the First Nuclear Age. The US is complementing this posture of stockpile maintenance with initiatives to develop missile defence and detection technologies to immunize itself against small nuclear attacks designed to destroy a city here or there, and is no longer preoccupied with deterring a massive attack designed to destroy the entire country, as during the Cold War. In a way that illustrates how scientific practices can change dramatically with shifts in the political environment, this change has been accompanied by a shift in the practice of weapons science in favour of simulations. While the strategic relationship with the Soviets was construed in terms of game theory, emphasizing the rationality of the enemy, America's new enemies are seen through the prism of a neo-colonial discourse that foregrounds their potential for irrationality and, therefore, undeterrability.

To return to the discussion at the beginning of the paper, a key difference between the first and Second Nuclear Ages, as Paul Bracken observes, concerns the larger number of players in the Second Nuclear Age, the disappearance of the binary superpower conflict, and the emergence of proliferation, both to smaller states and to sub-national actors, as a central issue in the system. The American national security state has been quick to recognize the aporia between the first and Second Nuclear Ages

Table 6.1 Contrasts between First and Second Nuclear Ages

	First Nuclear Age	*Second Nuclear Age*
Locus of conflict	Within core	Core-periphery periphery-periphery
Geopolitics	Bipolarity Nation-state	Globalization War on terror
Nature of contest	Qualitative arms race Quantitative arms race	Stasis in core Regional arms races Defence and detection
Strategy	Deterrence	Deterrence Pre-emption Missile defence
Enemy	Soviets	'Rogue states' Terrorists
Intellectual framework	Rationality Game theory	Post-rationality Neo-colonialism
Experiment	Nuclear tests	Simulations

and to remake its policies to take account of new threats and possibilities. Under the Clinton Administration we saw a (neo)-liberal strategy to exercise nuclear hegemony in the international system; under George W. Bush we are seeing an imperial strategy. We should not lose sight of the fact that both were strategies for hegemony.

Clinton's overarching goal was to create a liberal international order that would facilitate the flow of goods and services in a globalized economy while controlling the underside of globalization in the form of a global black market for nuclear materials. He also sought to isolate 'rogue states' from international civil society. Under Clinton the weapons labs were quick to absorb their former rivals in Russia into a co-operative clientilistic relationship and to develop global surveillance programmes to interdict the flow of nuclear technology to states conceived through an orientalist lens as insufficiently mature to wield nuclear weapons. While the right wing in the weapons labs wanted to break out of the test ban regime with a new generation of mini-nukes optimized for a new global struggle against rogue states, Clinton held back on this for fear of the damage it would inflict on his vision of a liberal global order. Ever the genius at triangulation, Clinton found a way to surrender nuclear testing to the international community without giving up American nuclear domination and a way to take testing away from the weapons labs while restoring their livelihood. By combining the test ban treaty with the Science Based Stockpile Stewardship programme, he sought to turn a test ban from an arms control initiative into a strategic firebreak between the advanced information economy of the US and the industrial economies of possible nuclear challengers. Stockpile stewardship helps bring nuclear weapons deeper into the information age so that, increasingly, the essence of nuclear weapons is seen, like DNA, to consist of code. It is only in the context of this conceptualization that we can, for example, understand the extraordinary hysteria over allegations that the Chinese-American scientist Wen Ho Lee shared with China discs containing bomb simulation codes.

By contrast with Bill Clinton, George W. Bush is on record opposing the Comprehensive Test Ban Treaty and his diplomats helped cause the collapse of the 2005 NonProliferation Treaty review conference by denying that the US had any obligations under Article VI of the treaty while seeking to block Iran's access to nuclear energy technology, although this access is guaranteed by Article IV of the treaty. (Article IV states that 'All the Parties to the Treaty undertake to facilitate, and have the right to participate in, the fullest possible exchange of equipment, materials and scientific and technological information for the peaceful uses of nuclear energy.') George W. Bush favours a unilateralist strategy to staunch nuclear proliferation either by means of pre-emptive strikes on proliferators or coercive diplomacy against them rather than through the international regimes of negotiation, surveillance and interdiction favoured by Clinton. Bush's vision is not of a liberal international order dominated by the US but of an American international imperium.

The position of Tony Blair, in his youth a supporter of CND, seems to have evolved alongside the American position. In the 1990s, as the Clinton Administration ramped up funding for stockpile stewardship and cut its nuclear stockpile, the Blair government followed suit.[22] Now it is rumoured that Blair plans to approve the design of a new generation of Trident warheads for the British stockpile. In any case he has approved an increase of the budget for British nuclear weapons research and development from its current level of about £300 million a year to £750 million by 2008 (McSmith, 2005).

It seems to be a rule that, among the many ways nuclear weapons corrupt those who develop them, they turn their owners into hypocrites. The hypocrisy of the Cold War was at least fairly transparent, as each superpower claimed that its own genocidal weapons were vital and virtuous, while the almost identical genocidal weapons of the enemy were hideous and evil. The hypocrisy of the Second Nuclear Age is less transparent and therefore in need of greater demystification. Western leaders, interpreting the world through an orientalist lens, behave as if the nuclear weapons of the advanced industrial powers are part of the natural order of things while the attempt by some developing countries to develop one ten-thousandth the destructive power currently wielded by the US represents a calamitous threat to global civilization. Indeed, to listen to some of the official rhetoric around the recent war in Iraq, one would have thought that Iraq was trying to acquire uniquely evil weapons, not weapons that the US and the UK already possess in abundance. This nuclear orientalist ideology, as old as colonialism and as new as the information age, is a fine place for anthropologists to enter the nuclear debate.

The First Nuclear Age produced two great intellectuals, one an American psychologist and the other a British historian. I am referring here to Robert Jay Lifton and E. P. Thompson.[23] Lifton drew on social and humanistic psychology to build a theory that analogized 'nuclearism' with the psychology of Nazi Germany in explaining how apparently decent people could work on genocidal weapons and how the American public could avert their gaze from what was being done in their name and to the very real possibility of their own extinction as a result. Lifton's model of 'nuclearism' foregrounds such phenomena as 'denial', 'psychic numbing', 'projection', and 'psychological doubling'.[24] E. P. Thompson, a Marxist who thought historically about structure and agency, wrote from the vantage point of Britain, a country that was simultaneously complicit with and marginal to the superpower arms race. In articulating his theory of 'exterminism', Thompson modified the classic Marxist narrative to theorize a nuclear mode of production that transcended structural differences between capitalist and Soviet society and produced a particular form of life characterized by heavy dependence on military spending and mass ideological subservience to mendacious government propaganda.[25]

Have Lifton and Thompson exhausted everything to be said about the nuclear age? One of the great polemicists of the Second Nuclear Age, Arundhati Roy (2000: xix), has said 'there's nothing new or original left to be said about nuclear weapons.' In this Roy is, unusually, wrong. While psychological mechanisms of denial and projection and economic structures of dependence on military spending may carry over from the First Nuclear Age, the Second Nuclear Age is unfolding in a context where the information sciences are remaking the phenomenology of weapons science, where globalization has replaced bipolarity and is reorganizing the political economy of nuclearism, where new nuclear actors are reshaping the dynamics of nuclear rivalry while the threat of further proliferation, possibly to non-state entities, is generating new ideological tropes of threat and security. This emergent nuclear age awaits its Lifton or its Thompson to critically articulate its deformative features. That person will not be Paul Bracken or any of the other intellectual lackeys of the American military-industrial complex producing rationalizations of American nuclear policy under the guise of doing serious intellectual work. Maybe the new articulation will, this time, arise from outside the West. Let us hope that, unlike the First Nuclear Age, this time anthropologists will play a role in the process of diagnosis and critique.

Notes

1. My thanks to Raminder Kaur for the original stimulus for this chapter, and to Achin Vanaik for his thoughtful responses to the chapter.
2. The few anthropologists to engage the Cold War or the nuclear age include Borneman (1998), Masco (2004, 2006), Rubinstein and Foster (1988, 1989) and Turner and Pitt (1989).
3. A notable recent exception is Lutz (2001).
4. Schwartz (1998). The number is in 1996 dollars.
5. For anthropological work on the Russian nuclear weapons complex, see Garb (1997). Komarova (2000), and Dalton et al. (1999).
6. For the work this research produced see Gusterson (1996, 2004).
7. See also Payne (1996) and Gray (1999). Left perspectives on the 'Second Nuclear Age' have been offered by Englehardt (2005) and Schell (2000).
8. On the manufacture of the term 'rogue states' see Klare (1995).
9. http://www.peacenowar.net/feb%2018%2002--news.htm.
10. Lyrics courtesy of Dr Ptolemy Schwartz.
11. A useful overview of the DARHT can be found at http://www.ligo.caltech.edu/LIGO_web/seminars/pdf/wilkinson.pdf. The UK has a functionally equivalent facility called Moguls.
12. See online minutes of a 1999 Aldermaston meeting at http://www.awe.co.uk/Images/LCM18_tcm6-2008.PDF.
13. The homepage of the French facility is http://www-lmj.cea.fr/html/cea.htm. A good but quite technical overview of the facility in English is given at http://ej.iop.org/links/q95/5K,0o4kS2fm8pbenxVFSsA/ppcf4_12B_012.pdf.
14. A good overview of the subcritical testing programme can be found at http://www.llnl.gov/str/Conrad.html. On the UK's participation, see http://www.yorkshirecnd.org.uk/news/articles/ukustest.htm.
15. The official Web site of this programme, now renamed ASC, is http://www.sandia.gov/NNSA/ASC/.
16. See Sigal (1997) for the best account of this crisis.
17. For histories of India's nuclear weapons programme, see Abraham (1998) and Perkovitch (1999).
18. Address to International Student Pugwash Conference, San Diego, June 1999.
19. For an anthropological account of the insecurity of Russian scientists in general during the 1990s, see Ninetto (2001).
20. For a critique of the suspect category, 'weapons of mass destruction', see Macfarlane (2005).
21. Excerpts from the Nuclear Posture Review can be read at http://www.globalsecurity.org/wmd/library/policy/dod/npr.htm. The draft doctrine for Joint Nuclear Operations can be accessed at http://www.globalsecurity.org/wmd/library/policy/dod/jp3_12fc2.pdf (accessed 14 November 2005). The document posted there is a draft and may differ from the final version.

22. Blair retired nuclear bombs formerly carried on Britain's Tornado aircraft and cut the British stockpile to about 200 weapons. For details of Britain's nuclear weapons programme, see http://www.nrdc.org/nuclear/nudb/datab18.asp.
23. Some will protest the exclusion of Jonathan Schell from this list. While Schell's books were written with extraordinary eloquence and insight, they were works of literary journalism that did not theorize the nuclear age in the way that Lifton and Thompson have.
24. See Lifton (1983), Lifton and Falk (1982), and Lifton and Markusen (1990).
25. See Thompson (1981, 1982, 1985).

Bibliography

Abraham, I. (1998), *The Making of the Indian Atomic Bomb: Science, Secrecy, and the Postcolonial State,* New York: Zed Books.

Borneman, J. (1998), *Subversions of International Order: Studies in the Political Anthropology of Culture,* Albany: SUNY Press.

Bracken, Paul (2003), 'The Structure of the Second Nuclear Age', *Orbis: A Journal of World Affairs*, 47(3): 399–413.

Dalton, R., Garb, P., Lovrich, N. and Whiteley, J. (1999), *Critical Masses: Citizens, Nuclear Weapons Production, and Environmental Destruction in the United States and Russia,* Cambridge, MA: MIT Press.

Englehardt, T. (2005), 'Our New Nuclear Age', *Mother Jones*, May 5. Available at http://www.motherjones.com/news/dailymojo/2005/05/our_nuclear_age.html.

Foster, M. L. and Rubinstein, R. A. (eds) (1986), *Peace and War: Cross-Cultural Perspectives*. New Brunswick, NJ: Transaction Books.

Garb, P. (1997), 'Complex Problems and No Clear Solutions: Difficulties of Defining and Assigning Culpability for Radiation Victimization in the Chelyabinsk Region of Russia', in B. R. Johnston (ed.), *Life and Death Matters: Human Rights at the End of the Millennium,* Walnut Creek: Altamira Press.

Gray, C. (1999), *The Second Nuclear Age,* Boulder, CO: Lynne Rienner.

Gusterson, H. (1996), *Nuclear Rites: A Weapons Laboratory at the End of the Cold War,* Berkeley, CA: University of California Press.

—— (1997), 'Studying Up Revisited', *Political and Legal Anthropology Review* 20(1): 114–19.

—— (2004), *People of the Bomb: Portraits of America's Nuclear Complex,* Minneapolis: University of Minnesota Press.

Johnson, R. (1996), 'The In-comprehensive Test Ban', *Bulletin of Atomic Scientists* 52(6): 30–5.

Kitfield, J. (1998), 'Fallout', *Government Executive,* 1 May 1998. Available at http://www.govexec.com/features/0598s1.htm.

Klare, M. (1995), *Rogue States and Nuclear Outlaws: America's Search for a New Foreign Policy,* New York: Hill & Wang.

Komarova, G. (2000), 'Ethnic Behavior Under Conditions of High Radiation', *Inner Asia,* 2: 63–72.

Krasniewicz, L. (1992), *Nuclear Summer: The Clash of Communities at the Seneca Women's Peace Encampment,* Ithaca, NY: Cornell University Press.

Lifton, R. J. (1983), *The Broken Connection: On Death and the Continuity of Life*, New York: Basic Books, Harper Colophone.

Lifton, R. J. and Falk, R. (1982), *Indefensible Weapons: The Political and Psychological Case Against Nuclearism,* New York: Basic Books, Harper Colophone.

Lifton, R. J. and Markusen, E. (1990), *The Genocidal Mentality: Nazi Holocaust and Nuclear Threat,* New York: Basic Books.

Lutz, C. (2001), *Homefront: A Military City and the American Twentieth Century,* Boston, MA: Beacon Press.

Macfarlane, A. (2005), 'All Weapons of Mass Destruction Are Not Equal', in MIT Center for International Studies (ed.), *Audit of Conventional Wisdom.* Available at http://web.mit.edu/cis/pdf/Audit_6_05_Macfarlane.pdf.

Marcus, G. (1995), 'Ethnography in/of the World System: The Emergence of Multi-Sited Ethnography', *Annual Review of Anthropology,* 24: 95–117.

Masco, J. (2004), 'Nuclear Technoaesthetics: Sensory Politics from Trinity to the Virtual Bomb at Los Alamos', *American Ethnologist* 31(3): 1–25.

—— (2006), *The Nuclear Borderlands.* Princeton: Princeton University Press.

May, M. and Haldeman, Z. (2004), 'Effectiveness of Nuclear Weapons against Buried Biological Agents', *Science and Global Security,* 12(1–2): 91–114.

McSmith, A. (2005), 'Blair's Nuclear Bombshell', *Independent,* 17 October.

Nader, L. (1997), 'The Phantom Factor: Impact of the Cold War on Anthropology', in N. Chomsky, R. C. Lewontin, H. Zinn, R. Ohmann, L. Nader, R. Siever, I. Wallerstein, I. Katznelson (eds), *The Cold War and the University: Toward an Intellectual History of the Postwar Years,* New York: New Press.

Nelson, R. W. (2002), 'Low-Yield Earth-Penetrating Nuclear Weapons', *Science and Global Security,* 10(1): 1–20.

—— (2004), 'Nuclear "Bunker Busters" Would More Likely Disperse than Destroy Buried Stockpiles of Biological and Chemical Agents', *Science and Global Security,* 12(1–2): 69–89.

Ninetto, A. (2001), 'Civilization and Its Insecurities: Traveling Scientists, Global Science, and National Progress in the Novosibirsk Akademgorok', in R. Stryker and J. Patico (eds), *Paradoxes of Progress: Globalization and Postsocialist Cultures,* Kroeber Anthropological Society Papers, Vol. 86, pp. 181–202, Berkeley: University of California.

Payne, K. B. (1996), *Deterrence in the Second Nuclear Age,* Lexington, KY: University Press of Kentucky.

Perkovitch, G. (1999), *India's Nuclear Bomb: The Impact on Global Proliferation,* Berkeley: University of California Press.

Price, D. (2004), *Threatening Anthropology: McCarthyism and the FBI's Surveillance of Activist Anthropologists.* Durham, NC: Duke University Press.

Roy, A. (2000), 'Introduction', in P. Bidwai and A. Vanaik (eds), *New Nukes: India, Pakistan and Global Nuclear Disarmament,* New York: Interlink Books.

Rubinstein, R. A. and Foster, M. L. (eds) (1988), *The Social Dynamics of Peace and Conflict: Culture in International Security,* Boulder, CO: Westview Press.

Schell, J. (2000), 'The Second Age of Nuclear Danger', *Boston Review: A Political and Literary Forum*, April/May.

Schwartz, S. (1998), *Atomic Audit: The Costs and Consequences of US Nuclear Weapons Since 1940*, Washington, DC: Brookings Institution Press.

Shroyer, J. A. (1997), *Secret Mesa: Inside the Los Alamos National Laboratory*, New York: Jonathan Wiley.

Sigal, Leon (1997), *Disarming Strangers: Nuclear Diplomacy With North Korea.* Princeton, NJ: Princeton University Press.

Taubes, G. (1995), 'Cold War Rivals Find Common Ground', *Science,* 268: 448–91.

Thompson, E. P. (1981), 'Letter to America', in E. P. Thompson and D. Smith (eds), *Protest and Survive: Stop Nuclear War*, New York: Monthly Review Press.

—— (1982), *Beyond the Cold War: A New Approach to the Arms Race and Nuclear Annihilation*, New York: Pantheon Books.

—— (1985), *The Heavy Dancers: Writings on War, Past and Future,* New York: Pantheon Books.

Turner, Paul and David Pitt (1989), *The Anthropology of War and Peace: Perspectives on the Nuclear Age*, South Hadley, MA: Begin & Garvey.

–7–

Genealogical Hybridities*

The Making and Unmaking of Blood Relatives and Predictive Knowledge in Breast Cancer Genetics

Sahra Gibbon

> Where the process of purification takes place relatively smoothly – where silence resounds about any given innovation – this too is fertile ground for social scientists. In this instance, the initial task is of course to name the hybrid, for it will usually be camouflaged as though it is a natural entity.
>
> Lock, 'On Dying Twice'

Introduction

Following the identification of two 'inherited susceptibility genes' in the mid-1990s (BRCA 1 and BRCA 2) the emergence of predictive genetic testing for an increased risk of breast cancer has been one of the most publicized and high-profile medical applications of a raft of recent rapid advances in human genetics since the mid-1990s. Such knowledge has meant that testing for these genes has now become available through specialist clinics in the NHS for those at 'high risk' of developing the disease and risk assessment is at the vanguard of an expanding field of clinical genetics (Wonderling et al., 2001).[1] This chapter examines how the arena of predictive medicine associated with breast cancer genetics is situated in relation to and in terms of different discourses and practices associated with genealogy, kinship and the family and the ways that these seemingly separate domains shape and inform one another. Examining how a particular kind of 'born-and-bred' kinship (Edwards, 2000) operates in practice at the interface between patients and practitioners, it points to the collective work at stake in the pursuit of predictive genetic knowledge. At the same time that the 'combinatory' power of such kinship thinking is helping to mobilize genetic knowledge or technologies

* This chapter was amended from *Breast Cancer Genes and the Gendering of Knowledge. Science and Citizenship in the Cultural Context of the 'New' Genetics*, by Sahra Gibbon, to be published in 2007. Reproduced by permission of Palgrave Macmillan.

and sustain the moral import and significance of 'blood' relations, this chapter also points to more disjunctured and uneasy implications for patients and practitioners. Identifying these gaps illustrates the hidden 'hybrids' (Lock, 2000) in what appear to be newly productive alignments between kinship and genetic medicine. This raises new questions about the powerful, but not always welcome, ricochet effects brought about by these conjunctions. It also suggests that social scientists must be attentive to the ambiguous and uneven terrains of kinship and genetic medicine in which genealogical knowledge is being 'put to work' (Edwards, 2000; Konrad, 2005; Gibbon, 2007).

The consequences of genetic knowledge for kin relations and families form an area of inquiry that social scientists have begun to explore. This includes examining the diverse ways that individuals and kin make sense of and incorporate or reject predictive knowledge in relation to new technologies such as genetic testing. Much of this work draws attention to the gaps that exist between kinship thinking or practices as lived experiences and the genealogical thinking that is assumed to be part of and reproduced by different kinds of genetic information (Marteau and Richards, 1996). The focus of interest is not only on the contrast between the 'powerful ethnoscience' of human genetics with 'kinship reckoning', which speaks 'a related and alternative language of possibility and truth' (Rapp, 1999) but also on the ability of the former to usurp the lived practices of the latter. At the same time others suggest that the medicalization or geneticization of kinship and the family is facilitated by the presumed 'fit' between the 'ideology' of the family and genetic inheritance (Finkler, 2001).

Examining the dynamics of clinical encounters in the context of clinical breast-cancer genetics points to more complex, flexible and ambivalent configurations of genealogical connection on both sides of the lay/practitioner divide. This raises questions about how, by whom and why kinship might be said to be increasingly 'geneticized' in an era of predictive medicine. In this chapter I draw on work that has examined how a particular kind of kinship thinking, which Edwards (2000) terms 'born and bred', is deployed by both patients and practitioners. Her ethnographic study of relatedness points to how ideas of what constitutes the seemingly fixed basis of kinship, for example 'blood', and what is more changeable or mutable, for example 'nurturance', is 'put to work' in a range of ways that traverse the seemingly fixed boundaries between the natural and the social (see also Strathern, 1992; Franklin, 2003; Nash, 2005; Sender, 2005; Wade, 2005). Ethnographic examination of clinical encounters suggest that a similarly fluid mobilization of ideas around genetic inheritance and the family is used, often in unexpected or contradictory ways, in pursuit of related but different ends by patients and practitioners.

The hybridity of the particular genealogical connections reproduced and sustained in this context operates at many levels. In one way this is evident in the first part of this chapter, which examines how the different activities of patients *and* practitioners help to reproduce and sustain genealogical relatedness. This 'co-production' of kinship suggests a field of action in which the locus and agency of geneticization are diversely and complexly embedded through the indexical power of 'substance' and 'code'. Nevertheless the work of translation and transmission at the clinical interface mobilized through and by the conjunctions between genealogy and genetics has uneasy consequences for patients (Bourret, 2002; Konrad, 2005) and also for practitioners. This chapter explores the tension generated between the individualism rooted in the desire to take pre-emptive action for those (mostly women in this case) seeking health interventions or codes of medical ethics orientated to care of 'the patient' and the more relational requirements of genetic information. These points of friction are often exacerbated by a gendered health agenda that informs the field of breast cancer and the ongoing uncertainty and technological limits that characterize this field of genetic medicine. Focusing on the interactional space of the clinic illustrates how both 'blood relatives' and predictive knowledge are being powerfully reproduced in breast-cancer genetics in disorientating and sometimes uneasy ways (Lock, 2005; Palladino, 2002; Gibbon, 2007).

Mobilizing Substance and Code; the Work of Co-production

The first part of this chapter, building on previous work (Gibbon, 2002), examines how the alignments between substance and code are reproduced and sustained by the work of patients and practitioners. Talking to patients I met about why they were attending the clinics and their expectations around forthcoming appointments revealed a particular reading and doing of family history before they were even seen in the clinic.

For many of those I met, the family history they were aware of pointed to an undeniable danger. This was exemplified in what Shona said:

> It obviously is genetic to some extent... It's just there is breast cancer down the female line on my mother's family and it just strikes me that it must be related unless it's just a coincidence that my grandmother and mother had breast cancer... But they had very different lives because one lived in the country and did nothing for years and years and the other lived in the city and just worked. So it's more probable that it's genetic that would be the logical explanation.

In general when there were several cases of breast cancer in the family, many of those I talked to were inclined to think that it was unlikely to be just a 'coincidence'. Genes provided an comprehensible, plausible and rational explanation for the cases of cancer in the family. This was brought home during my meeting with Donna after she talked about and then, after prompting from me, drew her family history.

Donna: It's so weird my Nan's sister is still alive, she was ninety last year. She's got all me mum's cousins they're all fine. But if you look at the tree there are two sisters, like one sister has it and my mum, who's the daughter and then the other line nothing.

Sahra: I wonder would you mind drawing how you see your family history?

Donna: No I don't mind. Going on me dad's side, I don't think of me dad's side as related to cancer funnily enough, because my Nan [on her father's side] did have stomach cancer and my granddad did have like a brain cancer but because the daughters didn't I don't seem to relate to that side. I just see like on my mum's side there is aunt kit and there is my Nan and then my mum [has had breast cancer] just like my Nan. Do you see what I mean [pointing to the diagram] see cancer and cancer we're scared that it's going to come. See if there is a certain gene it could have missed my aunt but went to my Nan.

Donna's very visual description of her family history helped to give form to a suspicion that genes were responsible for cancer in her family, a rationale that was reflected (and to a certain extent reproduced)[2] by the drawing I asked her to do (see Figure 7.1).

Donna divided this depiction of her family history into two halves, as her description had suggested. This started at the juncture between her grandmother and her grandmother's sister. The linear trajectory she describes and the danger she feels that this poses to her is clearly visible in the way she indicates who has had illness or cancer in the family. After this encounter I asked several other women to draw how they 'saw' their family history, copies of which are reproduced in Figure 7.2.

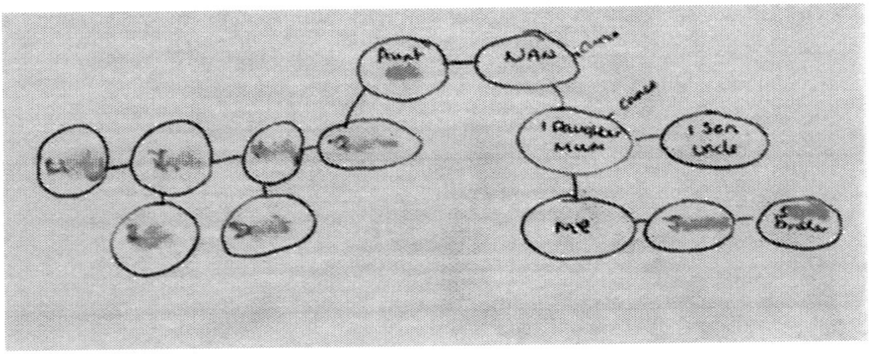

Figure 7.1 Donna's drawing

As these examples illustrate, it is striking that many of the resulting diagrams were dramatically pared down and somewhat pathologized representations of family history. In nearly all cases very few affinal relations were included or in some cases other consanguineous or 'blood' relatives. This was particularly so in Jane's and Shona's case where hardly any other relatives were drawn who might detract from a narrative trajectory of risk they had talked about. Jane drew only

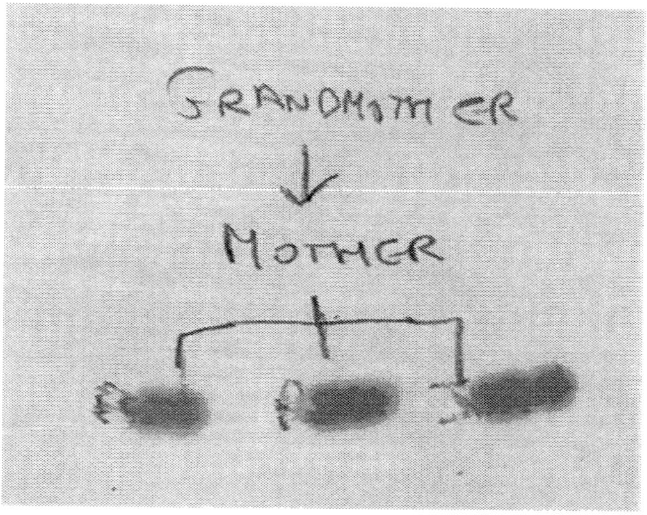

Figure 7.2(a) Jane's family history

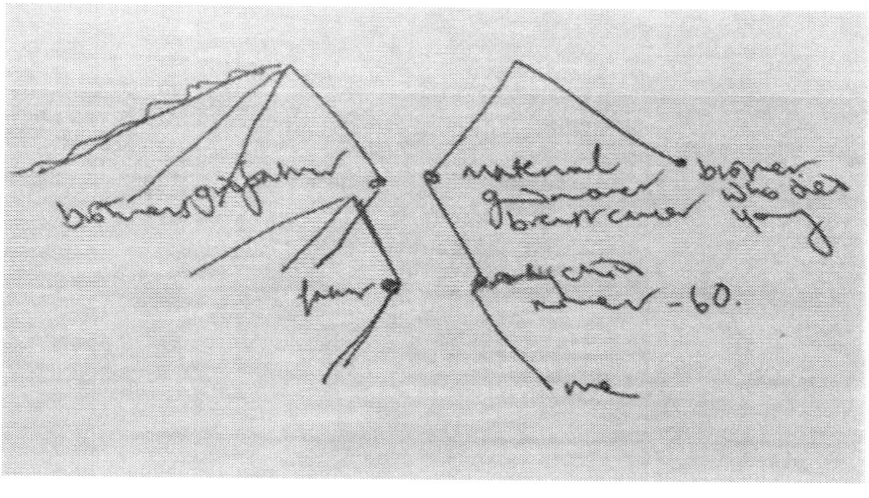

Figure 7.2(b) Shona's family history

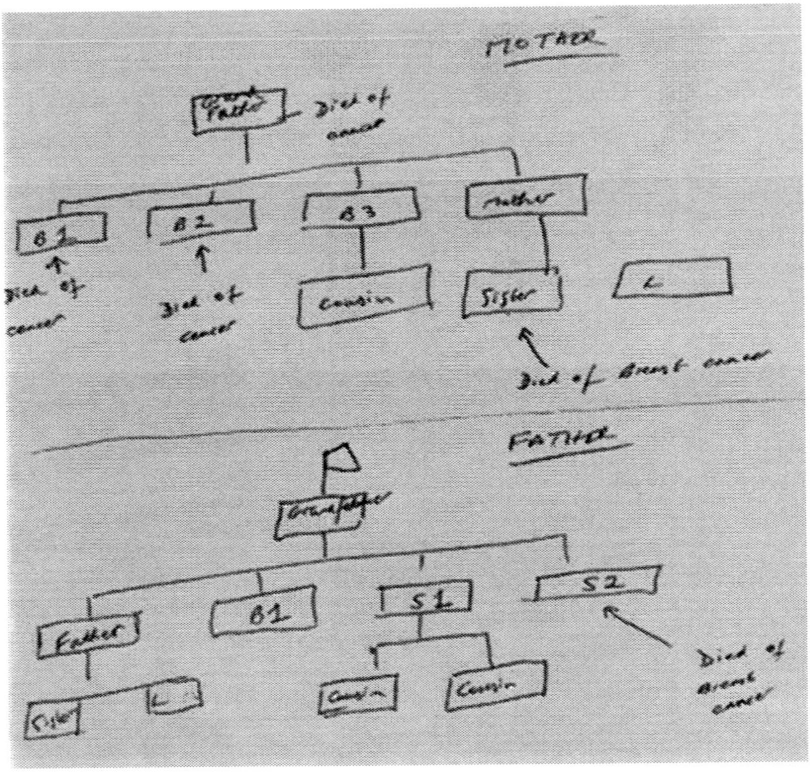

Figure 7.2(c) Lesley's family history

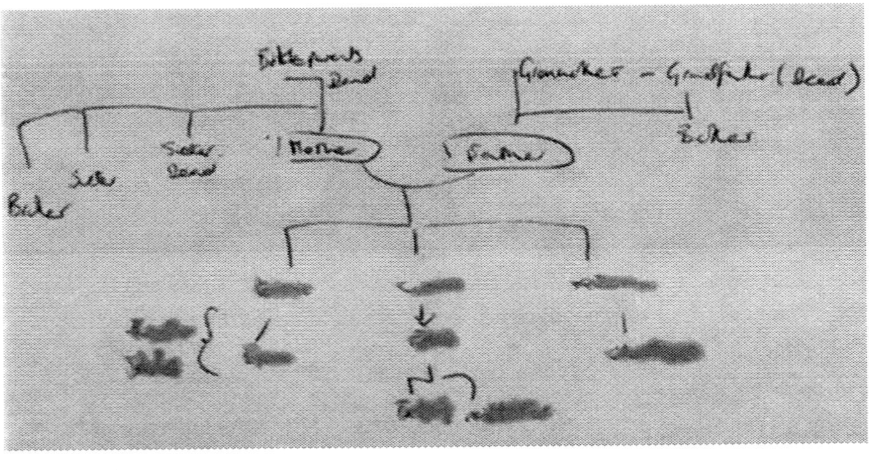

Figure 7.2(d) Lucy's family history

the two relatives affected by cancer (her mother and grandmother), herself and her two sisters. It was only much later in the interview that she mentioned in passing that she had four brothers. Even when the depiction of family history seemed to be more balanced, or at least to encompass more people, as in Leslie's case, 'clues' about possible risk were also visually represented. Lucy's depiction of her family history was less obviously a narrative about 'risk' in the way the others seemed to be. There was, for instance, no indication in her drawing of who had had cancer. However, unlike others she had also included her children and her sister's children in her depiction of family history.

These sparse and somewhat de-contextualized representations of genetic risk were also accompanied and to a certain extent re-inforced by the bodily parities and sharing of physical symptoms, build or even personality or life experience, which for some of these individuals provided further evidence of an underlying narrative of causation that strongly suggested the involvement of genes in some way (see also Richards and Hallowell, 1997). This was illustrated, for instance, in the way that, for a number of women whom I met, the sex of their children (and those of other family members) pointed to the way 'fate' and 'genes' were operating in interconnected ways. In one case, while talking about her family history and referring to the diagram that she had drawn, Lucy highlighted the fact that she, like her sister, had had girls. For her this provided an indication that there might be a genetic risk in the family.

> *Lucy*: These are all my sisters' here and they've nearly all got girls [pointing to the family tree she has drawn]. There is only one boy, which is another thing I think oh you know maybe that's connected to all this. My mum's sister had two girls and the aunt that died had got three girls, so not many boys in our family but lots of girls which does make you think...

As the diagrams outlined here suggest, family trees have a familiar iconic status and meaning for many outside a clinical setting. Tracing family history has in fact become a hugely popular activity and is part of a widespread commodity industry in a Euro-American context (Basu, 2006). However the evidence presented here suggests that it is not simply that patients' 'interest in genealogy' is connected to the need to attend to medical family history in the context of clinical appointments (Finkler, 2001: 240) but that genealogy, rendered in particularly geneticized ways, is actively sustained in how patients perceive risk and articulate family history. Rather than the notion of a 'positive family history' being 'an inversion which attempts the work of suppression' (Rapp, 1995: 130), here substantial efforts appears to be undertaken by patients to achieve such just such a pathologized and clinical classification.

But these articulations of geneticized relatedness must also be understood in terms of the pursuit by many patients of very specific ends where genes are inextricably linked to 'being' and 'becoming' (Duden, 2003). That is geneticized genealogies, at the same time as articulating deeply felt anxiety about the risk of developing breast cancer, also become a resource in seeking 'at-risk' patient status that will, it is hoped, secure ongoing care and the promise of what are perceived by some as preventative technological interventions. The way that evidence of risk is entangled with knowing and representing one's family history in the hope of preventing ill health in the future is also a process that cannot be abstracted from the gendered values and idioms of the increasingly high public profile of a culture of breast cancer activism, where this is a strong emphasis on the importance of health awareness and pre-emptive action (Gibbon, 2006, 2007).

But if patients are engaged in producing particular 'geneticized' readings of family history and relatedness in pursuit of certain ends, examining the work of practitioners in the cancer genetic clinic also points to activities that put kinship to work in contrasting ways. A number of social scientists have begun to point out how a focus on 'the family' is consonant with genetic medicine (Armstrong, Michie and Marteau, 1998; Kenen, Ardern-Jones and Eeles, 2003; Featherstone et al., 2006). Moreover the notion of the 'extended patient' (Bourret, 2005), or 'dividual' person (Palsson and Harðardóttir, 2000) have been useful for conceptualizing how in predictive medicine patienthood is being reconstituted relationally. But predictive medicine also relies on certain engagements and expertise on the part of practitioners in dealing with the social context and dynamics of family relations. Although these are themes that I explore in more detail elsewhere (Gibbon, 2007), here I simply want to highlight one way in which genealogical relations are configured by practitioners in the predictive practices of the clinic.

Producing the clinical family tree is central to risk assessment in breast cancer genetics. These types of representations use specific icons as a shorthand for gender, illness and death in the family so that what is notable, when they are presented or drawn up at the start of the consultation, is the visual impact of a variety of shorthand symbols or icons representing gender and disease in the family, as Figure 7.3 illustrates.

The presentation and/or production of the tree is accompanied by the clinician asking the patient for verbal confirmation about the information written on the form he or she has filled in concerning the types of illness or age of a relative's death. These are questions that reflect the centrality of this tool in making accurate assessments of risk. In the frequent absence of widespread predictive testing the tree is often the only tool of risk assessment or prediction and the basis on which further interventions may be offered to the patient.

From this perspective these paper technologies grounded, as they at least visually seem to be, in a particular 'origin story' about reproduction, identity,

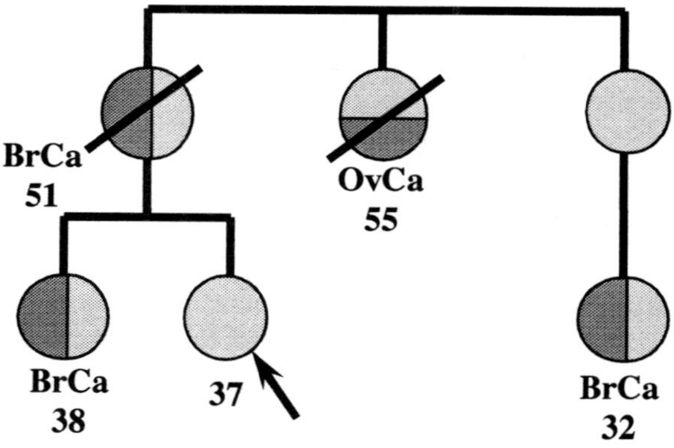

Figure 7.3 A clinical family tree

inheritance and risk, appear to articulate the power and credibility of scientific inference and knowledge (Rapp, 1999; Finkler, 2001). But this reading of family trees excludes the other ways in which genealogical connection and particularly the movement between 'born-and-bred' kinship is reproduced by practitioners in the space of the clinic. That is, the accuracy and the utility of risk assessments are reliant on practitioners understanding the social context and dynamics of family relations where hidden stories, myths, clues and/or family secrets and silences about the manner of illness or death are crucial to predictive practices (Gibbon, 2002). Moreover, in the absence of widespread genetic testing, clinical intervention depends on practitioners' abilities to facilitate, encourage and enable the sharing and mutual involvement of kin in each other's care. These 'pastoral skills' entail often detailed and delicate negotiations with different family members in securing mutual consent or agreement to undertake a certain procedure or further intervention such as genetic testing or the donation of a blood sample for research; skills that, although necessary, are not always easily or readily encompassed into clinical practice or seen as an obvious source of professional expertise (Gibbon, 2007).

The necessity of these skills illustrates how practitioners and patients put kinship and genealogy to work in diverse and often unexpected or contradictory ways in pursuit of different ends. For patients, a process akin to what Nahman (2005), in a different arena of high tech medicine, refers to as kind 'part/whole telescoping of traits' is apparent in the way that genealogical connection is articulated. In the space of the cancer genetic clinic, the past and future are condensed and ricochet back and forth in 'patients' desire to make risk real and apparent with the aim of securing an 'at-risk' identity that will enable ongoing care and monitoring. Although using at one level an ideology of genetic connection that articulates kinship as rooted in biology, the work that practitioners undertake with family trees sits alongside the need for skills in dealing with the social dynamics and context of the family. In this sense the 'co-production' of genealogical knowledge is far from obvious or stable, subject to and locus for traffic between the natural and the social that works in 'flexible' (Nash, 2005) or what Strathern (1992) terms 'merographic' ways. The fluidity of this double movement of a particular kind of kinship thinking, rather than being thought of as de-stabilizing, seems to reference the 'combinatory power of substance and code' (Carsten, 2001), enabling and facilitating genetic knowledge in this particular field of medicine.

Genealogical Hybridities and Predictive Uncertainty

In contrasting the relative silence that has surrounded the implementation of organ transplant technology in the US with the more open and obvious disquiet about such technology in Japan, Margaret Lock (2000) suggests that 'naming' the hybrid is an important task for social scientists, camouflaged as it might be as a 'natural entity'. The second part of this chapter examines how the constructions of genealogy in breast cancer genetics, although seemingly powerful, can in fact have more complex and uneasy consequences for different people. The detailed experiences and narratives of two women I met bring to the fore the tensions this generates for some patients. In both cases the penumbra of being a patient, in pursuit of care of the self (seen in terms of a desire for on-going monitoring and surveillance) was not always easily extended or gratefully received by their kin and, in fact, reverberated through families in uneasy ways.

Emily

During our first meeting Emily, a woman in her twenties completing her academic studies, talked about her forthcoming visit to the clinic as a means through which she could impart information and care to others in the family: 'If, with my scientific background, I can get the facts right then I can talk to all my cousins about it.' But she also pointed out that many family members did not at that moment want to be involved. This awareness was sharpened and given a more real hue after her visit, as Emily explained when we met for a second time:

> The doctor explained that there are various options. One would be having a genetic test so my mother would be tested to see if she's got a gene then to see if I've got it as well. The other one was that they could take my mother's blood and freeze it and keep it there so possibly in the future she could have a genetic test or if something really bad happened at least we would have a sample of her that we could test. So I thought all that

> was good and a brilliant suggestion. But I went home and spoke to my mother and she got in a right state. The way she feels is that she just wants to ignore the fact that it ever happened and she doesn't want to be tested for the gene she doesn't want to have blood taken. Now that puts me in a difficult position.

Despite this impediment to shared or future care Emily continued to identify as a patient in a relational sense; something that, for her, held out the hope that some intervention might be possible at a later date.

> *Emily*: My mother's got more sisters and if another one of them went down with it or if something happened to sort of like my nieces, then I think we'd have to seriously start thinking about investigating it. I think I'd want to go back to my mother and see if she'd donate blood. They [the clinic] said they've got new methods of testing coming out and if in the future maybe one of my cousins wants to get involved they [the clinic] said that's all cool ... which sort of means that my entire family would be well looked after.
>
> *Sahra*: Have you talked to any of them about doing that?
>
> *Emily*: I've spoken to a couple of my female cousins and they're sort of in complete denial. Because neither of them has got a mother who has been through it and so they're sort of 'oh no I'll be fine don't worry about it.' But the cousin of my mum's mum is also in complete denial, I don't think it's worth pursuing because she's not in the frame of mind to want to listen about it.

Emily's own care, even after her mother had declined to donate blood for storage, was still predicated therefore in part on the future health and involvement of others. Clearly Emily saw her involvement in the clinic in terms of 'caring' for her family, even if they continued to be more reluctant participants in this exchange of rights and obligations.

Jane

During our first meeting, Jane, who was in her early fifties, talked about how she had completed the family history form. Like many people attending the clinics, there were gaps in her knowledge of the family history, as her description of this process demonstrated

> Well I spoke to mum and dad obviously and brothers and sisters. On me mum's side, I don't know. On her father's side I know nothing, she doesn't even know who her father was. All she knows is that it was a married man in the 1920s but that's all. So there is no one else on her side of the family alive apart from some distant cousins who don't even live in this country any more.

However, following Jane's visit to the clinic, it was clear that this somewhat fuzzy detail concerning her mother's father had greater significance than she had thought. This was revealed in the first few moments of our second interview when I asked her to tell me about what had happened.

> *Jane*: well, everything is fine. I've only got a very low risk of a hereditary gene. Although they picked up, my mum, as I explained to you I think, was illegitimate. Well her father was a Russian Jew. I don't know anything about him only that he was Jewish and married ... mum sort of said over the years, we've all got Jewish blood in us. All these sort of jokes come out. Anyway I happened to mention to the nurse, she was asking if I knew anything on that side of the family, and I said well I don't know anything about that side of the family apart from the fact that he was a Russian Jewish immigrant. She said 'oh that's quite interesting because that puts a slightly different slant on the hereditary [aspect].' Apparently there is a slightly different gene that comes through on the Jewish side. So mum has agreed to have her DNA tested and she's given permission to have her medical notes to be looked at, to see if there is any link that way.

In the course of our discussion Jane talked about how the medical interest in this aspect of her family history had impacted on her and her family and the range of issues it had raised

> My mum thinks she knows the name [of her father] but she's not even a 100 per cent sure on the name. I think the general view of everybody is well we didn't know him, so it doesn't matter (laughing). All be it perhaps a little tiny bit of him is in all of us. Now that we know that perhaps there was something that he could have passed through it makes us feel that perhaps he's a little bit more important... So it's quite a wide thing it could open up. Suddenly there is a grandfather that has got a lot of importance in our lives, or could have a lot to do with things that happened to us. Everybody has said oh he was just someone who went with my grandmother, but he is a bit more important than just a 'fly by night' if you like.

It was evident, therefore, that this 'new thing' had opened up many issues. For instance the re-appearance of this figure from the past seemed to provide renewed focus and force for feelings of guilt or anxiety among different family members.[3] This sense of unease was apparent in her final comments when I asked Jane whether she felt overall her appointment had been a good thing to do, given that regular screening had been set up for her and other investigations were under way with her mother.

> *Jane*: Overall I would say that it's settled a lot. The only thing is it's thrown up the thing with the Jewish ancestry, that's the only thing. I'm hoping that it won't be years before we get an answer from that and that if it is something, a gene on that side of

> the family, if it's there then start dealing with that. It's a new thing that's come in and there are no answers to it. But you see because of the Jewish link, it worries me more, because it is more prevalent. [Reading from hospital letter] 'if we do find a fault in the blood sample and testing may make take many months or years and it would mean that anyone in the family who wished to know their genetic status could have a genetic test after counselling.' I do remember reading that there is something that affects the Jewish population, I think they get blisters. I seem to remember reading or hearing about some disease, it's a bit like black people get sickle-cell anaemia... So I wondered whether there is anything else ... it's just well hopefully we will get answers, because if we have to wait a year, you'll be worrying. It's the unknown, that's the only doubt now.

The new significance given to a shadowy figure from her grandmother's past had obviously reverberated through Jane's family. Although this might have given renewed vent to older issues, her comments also suggested that it had also raised new and troubling questions.

The case studies presented here show how unknowing, reluctant, or in Jane's case, essentially unknown relatives, became entangled in each other's medical care in ways that resituated risk, rights and responsibilities between related individuals. The responsible self-actualizing individualism of preventative health-care practices, which in part arises from and is also strongly propelled by gendered cultures of health activism and awareness (see Gibbon, 2006), can come into conflict with a necessarily more distributed action and shared investment required and demanded by predictive knowledge. For Emily, fulfilling the requirements for being a patient, in the hope of ongoing monitoring and future care, was thwarted somewhat by the reluctance of her mother to participate in her daughter's preventative goals. The significance of Jane's unknown relative from the past, unearthed in her own pursuit of health-care, raised old and new issues about risk, responsibility and the consequences of predictive knowledge for herself and others. When, as Konrad (2005: 152) points out, heredity in an era of genetic medicine becomes 'what one decides to tell and how knowledge is shared and made lively as social anatomies of interdependence', we can see that the kind of genealogical connections required by predictive practices do not align neatly with the pathways of shared genetic inheritance.

Yet the sorts of 'bi-directional affective entanglements' (Rose, 2001: 10) that arise from predictive health-care practices also encompass and implicate practitioners . They can find themselves having to balance the demands of a code of medical ethics orientated to the medical care of individuals with negotiating the often different demands, concerns and care of related members of a family. The difficulties that this creates are illustrated in the way one practitioner talked about the problems of pursuing predictive knowledge where the 'always present effect' (Strathern, 1992) of kin relations could compromise an ability to act expertly. For instance, this is what one clinician said when she pointed out the difficulties of dealing with family members who were being treated in different centres:

> I think the most difficult thing is, that I find it very hard to reconcile within myself, the issue we have on confidentiality between centres. I understand that patients have to have their confidentiality respected. You know the information has been given to you by one person and if you're passing it on to help the management of a family in another centre, then in theory then you should be doing nothing but good. I can see that genetics is different because you've got information about others, but we've got to get around this because otherwise something dangerous is going to happen. We're starting to have this battle between the rights to know of the people who are at risk versus the rights to confidentiality of those who already have at risk information.

As Konrad (2005) points out in relation to predictive knowledge for Huntingdon's disease, the line of connection between 'pre-emptive individualism' and collective patienthood, via the family is at the very least ambiguous and sometimes ambivalent and often difficult to negotiate for patients and I would suggest practitioners also. Such challenges are compounded here by the fact that this oscillation is refracted through an ethic of female nurturance and the 'complex' nature of predictive interventions in breast cancer genetics, with its accompanying technological uncertainty and limited treatment interventions.[4] Contingency is a central feature of this evolving predictive medical practice. The 'rub' in this and other arenas of predictive medicine is that 'there is no such thing as prognostic information per se, only different categories of information that generate different prognostic fields for the moments of potential knowledge disclosure' (Konrad, 2005: 84). The health care encounter recounted below illustrates how the challenges of dealing with 'divinatory' practices are in fact caught up with and compounded by the genealogical hybridities of breast cancer genetics. The appointment concerned a husband and wife in their late fifties who had been to the hospital on several occasions but who were coming for what was intended to be their final visit before making a difficult decision about further possible interventions:

> The woman sitting with her husband in the clinic has been diagnosed with breast cancer a few years ago and is now in remission, following chemotherapy and a mastectomy to remove one of her breasts. Before they come into the consulting room, the doctor tells me that they are here today to help them make a decision about whether or not to have a prophylactic mastectomy on the woman's healthy breast, given her family history.
>
> They arrive clearly prepared with a folder full of cuttings and questions already planned. After responding to their queries, the consultant proceeds to talk about genetic testing. Although it hasn't as yet been brought up by either of them, their response suggests that it's clearly an issue that they have been discussing in some depth as the woman explains

Patient: I think we've decided not to go ahead with the test. I mean I can quite happily walk around thinking that I am probably positive for the gene but if somebody was to say oh yes 100 per cent you are positive for the gene then I would find that difficult and also I've got to think of my daughter, she's only 22. I don't want to place too much burden on her and she's got insurance to think about as well.

There is slight pause before the patient asks the doctor in a rather worried way whether or not the hospital would only go ahead with surgery if she had a positive test. The consultant says no, they will still be willing to go ahead with it. This prompts the woman's husband to raise further questions about the value of testing.

Patient's husband: I'm wondering about this then so if she tests positive for the gene then that doesn't mean that she is necessarily going to get it and if she tests negative for the gene that doesn't mean that she hasn't got it as well, is that right?
Consultant: Yes that is right
Patient's husband: Well what is the point in that then!
Consultant: Well it would just be *neater* to have a genetic test because then if it was negative you really wouldn't have to worry so much and might then not want to go ahead with the surgery.

Making things 'neater' did not, as expected, convince the couple at the end of the appointment to go ahead with genetic testing. In this situation, where a woman had had breast cancer and did not see the benefit (but in fact identified the dangers) of predictive knowledge for her and her family the clinician seemed to be left floundering for appropriate reasons to undertake this procedure. Ultimately she resorted to an explanation that served to compound the lack of benefits that the couple had already highlighted. This exposed not only the comprised care giving engendered by predictive interventions but the also limits to practitioner expertise.

Conclusion

Examining how genealogical connection is reproduced and mobilized at the clinical interface in the pursuit of predictive knowledge illustrates the diverse means by which genes, kin and the family are put to work in this context. The consequences of such collective work are as unpredictable or ambivalent as they are powerful. As Palladino (2002: 140) points out, exploring the 'dialogical' spaces of the new genetics 'calls into question the stability of "knowledge"' and brings into view 'a more fissiparous and centrifugal discourse.' The work of patients in pursuing certain 'at-risk' identities reproduces a particularly geneticized reading of family history and relatedness – a practice that is intimately tied to the investment, hope and promise of predictive medicine. This is a mode of patient identification and anticipation directly linked to the culture of female activism and awareness that is so much a part of a public-health discourse of breast cancer (Gibbon, 2007; see also Klawiter, 2002). At the same time the demands of predictive knowledge in enlisting and requiring the mutual involvement and shared participation and engagement of related family members ensures that DNA, far from being 'devoid of morality or affect, the hallmark of family and kinship relations' (Finkler, 2000: 187), is actively imbued with just those qualities in the clinic. We have seen how this is sustained by the necessary enlisting of kin in acquiring blood samples or sharing of information in undertaking risk assessment. In this sense the transmission of knowledge and care in breast cancer genetics is embedded in and makes use of 'the combinatory power of substance and code ... at the heart of the notion of the blood relative' (Carsten, 2001: 50), productively connecting and collectively producing kinship and DNA.[5] What seems like a hybrid configuration is in fact a powerful articulation of mutually encompassing aspects of the domains of genetic science, gender and the family.[6] This suggests not opposition or disjuncture but newly productive alignments and where strategic mobilizations of both the natural and the social are undertaken in the dynamics of the clinic encounters and the collective pursuit of knowledge, although this may ultimately be for different ends.

The rapid translation and the relative ease with which breast cancer genetics has been translated into clinical practice suggests that any 'hybrids' which characterize this arena of health care are perhaps more hidden and must be sought elsewhere. Some of the gaps and disjunctures of this field of medicine have been brought to the fore by reflecting on the narratives of two women attending the clinics. Here the necessary and required 'oscillation' between the individual and family in predictive medicine (Bourret, 2005) has been shown to exacerbate the existing moral frameworks of obligation and exchange among kin relations with consequences for those attending the clinic and practitioners. However the genealogical hybridities of breast cancer genetic also arise not just because 'everyday kinship practices are as much about dividing and excluding as they are about connecting' (Strathern, 1996: 530) but because contingency, uncertainty and questions about the utility and viability of technological interventions such as genetic testing continue haunt this arena of medical practice.

In conclusion I want to illustrate how the border crossing activities between the social and natural identified in this paper are put to work in an attempt to suppress the fact of contingency by examining an exchange that took place between a patient and a practitioner at the end of one appointment that I observed:

> The woman sitting in the clinic is here on her first visit. It is a lengthy appointment, in part because a gene mutation has been identified in a relative and the doctor spends much time outlining the 'choices' available to her. There has also been some discussion

prompted by the patient about the limits of undertaking the procedures being offered to her. Nevertheless she concludes the appointment on a suitably upbeat note saying: 'your aunt choosing to have this genetic test, is a *legacy*. If she hadn't done that then we wouldn't be able to know for certain that you could be at risk, it explains things. The information is not going to go away and will always be here as a resource to use when you and your family want to' [my emphasis].

After the patient has left the doctor talks to me about how 'easy' and 'straightforward' the consultation was because 'when you know that there is a gene in the family you can offer definite things and provide them with real options'. However these comments seemed to sit in tension with the patient's less than enthusiastic response, for whom these options did not necessarily generate such easy or straightforward answers.

The clinician's choice of language in terms of a 'legacy' for future generations not only reveals the challenges for practitioners working in this setting, but how the morality and value of kin relations and predictive knowledge are complexly entwined in this field of medical practice. The example draws attention to the importance of examining not how genealogical connections are put to work in attempts to suture the gaps and lacunae that are characteristic aspects of clinical breast cancer genetics, but also the hidden hybridities of this and other emerging arenas of predictive medicine.

Notes

1. The field of genetic medicine and in particular cancer genetics has been an area of increased government investment over the last few years reflecting the hope that a so-called 'genetic revolution' will align prediction with prevention in supporting and facilitating targeted intervention with effective social and economic rationalization of increasingly finite health care services (see Milburn, 2001).
2. These drawings were undertaken some weeks prior to first-time appointments at the clinic. It is therefore possible that some of those I met perceived me as a 'gatekeeper', in terms of access to the clinic and therefore reproduced particular kind of drawings in relation to their expectations around forthcoming visits to the cancer genetic clinic. Nevertheless despite this possibility the drawings do index a revealing representation of family history.
3. There was evidence from our previous interview that a discourse of 'guilt' was present among family members as a result of another condition that both her husband and her son had, as Jane pointed out: 'We've already got problems with my husband whose got hypercholesterolemia and the sons got it as well and the chances are that he's passed it on. That causes its own problems and there is quite a sort of guilt trip because they've passed it on.'

4. There are, in fact, significant limits in relation to the predictive technologies of breast cancer genetics. The move towards routine predictive genetic testing for breast cancer in the US (Parthasarathy, 2005), France (Bourret, 2005) or for colon cancer in the UK (Palladino, 2002) has been far from assured or straightforward. In the UK predictive genetic testing for breast cancer is currently only generally available to those persons in a family where a gene mutation has been identified in a living affected relative or from whom a blood sample can be obtained. Moreover because both the BRCA genes are large, with hundreds of different mutations, looking for unknown mutations on the gene is time consuming and difficult. During my research this meant that it was only possible to test approximately 60 per cent of the BRCA1 gene, looking at 'hot spots' where mutations were thought to lie, and it was estimated that this picked up about 65 per cent of mutations. Once a mutation has been found then predictive genetic testing of other members of the family can be done relatively simply. But for large numbers of people referred or attending the clinics a conclusive test result is simply not possible. This may be because they have no living affected relatives, or because they were still waiting for a result, which could take many months (if not years in some cases) and/or because the result was inconclusive, linked to an inability to test for all the gene. Others may have received a 'negative' test from the BRCA1 gene and were waiting for a BRCA2 test result, which was very much in its infancy at this time. Moreover even for those identified as carrier of a gene mutation the estimated lifetime chance of developing breast cancer is uncertain, with the initial much touted figure of 80 per cent now subject to significant revision and re-assessment (Bourret, 2005).
5. As I discuss elsewhere, such alignments also have consequences for the way that particular articulations of female gender, especially in respect to ideas of nurturance are reproduced also (Gibbon, 2006).
6. See Wade (2005) for further discussion of the productivity of hybrid connections between the social and natural.

Bibliography

Armstrong, D., Michie, S. and Marteau, T. (1998), 'Revealed Identity: A Study of the Process of Genetic Counselling', *Social Science and Medicine*, 47(11): 1653–8.

Basu, P. (2006), *Highland Homecomings: Genealogy and Heritage Tourism in the Scottish Diaspora*, London: UCL Press.

Bourret, P. (2005), 'BRCA Patients and Clinical Collectives: New Configurations of Action in Cancer Genetic Practices', *Social Studies of Science*, 35(1): 41–68.

Carsten, J. (2001), 'Substantivism, Anti-substantivism, and Anti-anti-substantivism' in S. Franklin and S. McKinnon (eds), *Relative Values: New Directions in Kinship Studies,* Durham, NC: Duke University Press.

Duden, B.(2003), 'Keynote Address'. Paper presented at Genes, Gender and Generation Workshop, Lancaster University, June.

Edwards, J. (2000), *Born and Bred. Idioms of Kinship and New Reproductive Technologies in England,* Oxford: Oxford University Press.

Finkler, J. (2001), '"The Kin in the Gene": The Medicalization of Family and Kinship in American Society', *Current Anthropology*, 42(2): 235–63.

Featherstone, K., Atkinson, P. A. , Bharadwaj, A. and Clarke, A. J. (2006), *Risky Relations: Family and Kinship in the Era of New Genetics*, Oxford: Berg.

Franklin, S. (2003), 'Re-Thinking Nature-Culture: Anthropology and the New Genetics', *Anthropological Theory*, 3(1): 65–85.

Gibbon, S. (2002), 'Family trees in Clinical Cancer Genetics: Re-Examining Geneticisation', *Science as Culture*, 11(4): 429–57.

—— (2006), 'Nurturing Women and the BRCA Genes: Gender, Activism and the Paradox of Health Awareness', *Anthropology and Medicine* (in press).

—— (2007), *Breast Cancer Genes and The Gendering of Knowledge*, Basingstoke: Palgrave Macmillan.

Hallowell, N. and Richards, M. (1997), 'Understanding Life's Lottery: An Evaluation of Studies of Genetic Risk Awareness', *Journal of Health* Psychology, 2(1): 31–43.

Kenen, R., Ardern-Jones, A. and Eeles, R. (2003), 'Family stories and the Use of Heuristics: Women from Suspected Hereditary Breast and Ovarian Cancer Families', *Sociology of Health and Illness*, 25(7): 838–65.

Klawiter, M. (2002), 'Breast Cancer in Two Regimes: The Impact of Social Movements on Illness Experience', *Sociology of Health and Illness*, 26(6): 845–74.

Konrad, M. (2005), *Narrating the New Predictive Genetics: Ethics, Ethnography and Science,* Cambridge: Cambridge University Press.

Lock, M. (2000), 'On Dying Twice: Culture, Technology and the Determination of Death' in M. Lock, A. Young and A. Cambrosio (eds), *Living and Working with the New Medical Technologies: Intersections of Inquiry*, Cambridge, New York: Cambridge University Press.

—— (2005), 'The Eclipse of the Gene and the Return of Divination', *Current Anthropology*, 46 (Supplement): S47–S70.

Marteau, T. and Richards, M. (eds) (1996), *The Troubled Helix: Social and Psychological Implications of The New Human Genetics*, Cambridge: Cambridge University Press.

Milburn, A. (2001), Speech by the Secretary of State for Health at the Institute of Human Genetics, International Centre for Life, Newcastle Upon Tyne, 19 April.

Nahman, M. (2005), *Israeli Extraction: An Ethnographic Study of Egg Donation and* National Imaginaries, PhD thesis, Lancaster University.

Nash, C. (2005), Geographies of Relatedness, *Transactions of the Institute of British Geographers*, 30(4): 449–62.

Pálsson, G. and Harðardóttir, K. E. (2002), 'For Whom the Cell Tolls: Debates About Biomedicine', *Current Anthropology,* 43(2): 271–301.

Palladino, P. (2002), 'Between Knowledge and Practice: On Medical Professionals, Patients and the Making of the Genetics of Cancer', *Social Studies of Science*, 32(1): 137–65.

Parthasarathy, S. (2005), 'Reconceptualising Technology Transfer: The Challenge of Building an International System of Genetic Testing for Breast Cancer', in D. Guston and D. Sarewitz (eds), *Shaping Science and Technology Policy: The Next Generation of Research*, Madison, WI: University of Wisconsin Press.

Rapp, R. (1995), 'Constructing Amniocentesis: Maternal and Medical Discourses', in F. Ginsburg and R. Rapp (eds), *Conceiving the New World Order: The Global Politics of Reproduction*, Berkeley: University of California Press.

—— (1999) *Testing Women, Testing the Fetus: the Social Impact of Amniocentesis in America.* New York: Routledge.

Rose, N. (2001), 'The Politics of Life Itself', *Theory, Culture and Society*, 18(6): 1–30.

Sender, P. (2005), 'How Many Rolls in a Wheel? Genetics and Deterministic Theories', paper presented at Medical Anthropology Seminar, University College London, January.

Strathern, M. (1992), *After Nature: English Kinship in the Late Twentieth Century*, Cambridge: Cambridge University Press.

—— (1996), 'Cutting the network', *Journal of the Royal Anthropological Institute*, 2(3): 517–35.

Wade, P. (2005), 'Hybridity Theory and Kinship Thinking', *Cultural Studies*, 19(5): 602–21.

Wonderling, D., Hopwood, P., Cull, A., Douglas, F., Watson, M., Burn, J. and McPherson, K. (2001), 'A Descriptive Study of UK Cancer Genetics Services: An Emerging Clinical Response to the New Genetics', *British Journal of Cancer*, 85(2): 166–70.

–8–

Where Do We Find Our Monsters?

Debbora Battaglia

When I open to the first page of Donna Haraway's *Modest_Witness@Second_Millenium.FemaleMan©_Meets_OncoMouseTM,* I find that I have entered the future world of Marge Piercy's *He, She and It* (1991).[1] In this world 'a seventeenth-century golem in Prague's ghetto and a twenty-first-century cyborg in a Jewish freetown in North America are blasphemously brought into being to defend their endangered communities'. Living inside the hills in a society of women – all of whom remain in the wake of an apocalyptic war in the Middle East – Nili, the cyborg, describes their reproductive strategies:

> We clone and engineer genes. After birth we undergo additional alteration. We have created ourselves to endure, to survive, to hold our land ... *I am sent like the dove or maybe the raven from Noah's ark* to find out if the world is ready for us, and also if there's anything out there we might want. (Piercy, in Haraway, 1997: 1–2)

'Eve'

On 26 December 2002, the chemist Brigitte Boisselier, Director of Clonaid.com and a bishop in the Raelian religion, announced to an assembled press in Florida that she had cloned a baby girl, who was to be called Eve. Dr Boisselier produced no physical evidence – no DNA print, not even a photograph (normally the 'passport to the cosmos'[2] of a world-wide media audience). The raw claim was enough to send a groundswell of public and professional debate on reproductive cloning and by extension therapeutic cloning and assisted reproduction, across the global mediascape. Eve, it can be said, had become the first media-cloned child of a faith-based scientific programme.

The Raelians[3] are a creationist science religious movement that claims between 2,800 and 55,000 members world-wide, depending on your source, drawn to the idea that extraterrestrial scientists created life on Earth, in particular creating human beings 'in their own image' through advanced cloning technology. Galvanized by the charismatic presence of Rael, the French-born singer-songwriter, journalist, Formula I race car driver and prophet formerly known as

Claude Vorillhon, Raelians hold that human beings are the Designers' supreme 'artworks' in progress, evolving in accordance with natural law and in pace with technoscientific advances toward a time when genetic and robotic engineering will make a world safe for cyborgs and clones (Rael, 2005/[1998]).

Perhaps surprisingly, the movement attracts, among others, mainstream scientists whose scientific work is an act of faith, and whose spirituality is motivated by science. Showcased on the Rael.org Web site are testimonials from nanotechnologists, mechanical engineers, chemists, biologists, psychologists and my primary consultant, a neuroscientist who, during the period of my online and offline research between 2001 and 2003, held a postdoctoral appointment at Harvard. Many of these professionals hide their star pendants under their shirts when they go to the workplace; some have lost jobs or job prospects as a result of being Raelian. In part this is due to the prevailing perception in the scientific community (including transhumanists and bioethicists) that Raelian scientists are alarmingly misguided. And they are no less problematic from the viewpoint of organized religion. While they align themselves with intelligent design creationists – the so-called Neo-Creos of the Christian Right – Raelians are not only atheistic but actively progressive. The Raelian movement I know is basically a make-love-not-war peace movement. Pro-choice, pro-gay, and anti-war, Year One of the Raelian calendar is the bombing of Hiroshima. Yet even if we bring about a nuclear holocaust, the Designers will use the DNA of 'the awakened', scanned and archived while circling the planet in the arks of surveillance spaceships, to clone worthy human beings on a more hospitable planet. The 9/11 terrorist attacks and violence in the Middle East are taken as evidence that the end is likely, and a matter of urgent anthropology.

All of this may sound fantastical. However, the Raelians' baby Eve was taken as a real possibility – to the extent, for example, that a few weeks after the announcement of her birth, I could tune in to *Court TV* and watch as an attorney in Florida stated his case to remove the child from her parental guardians, based on her (still speculative) clonehood.[4]

Meanwhile, the birth announcement was unsettling my ethnographic fieldwork with Raelians in Canada, the US, and in multiple cyber-localities. My email screen and voice-messaging system were flooded with colleagues' inquiries; I was invited to dine with college administrators to discuss the case; friends were sending me front-page newspaper stories from their fieldsites in remote corners of India; reporters from the *Wall Street Journal* and the *New York Times* were taking a sudden interest in my work and requesting 'off-the-record' statements. Ordinary Raelians and even those in the administrative structure were asking if I knew anything that they didn't about the story. In short, the effect of a rushing community-on-the-flow was intense, as multiple agents cross-connected by accident or design at telecommunication access sites. I, meanwhile, remained ignorant of 'the truth' of the story until it no longer mattered to anyone (that is, until it was utterly press dead). Which was for the best. On the one hand, I felt at odds with a Raelian publicity machine that sought to make somatic cell nuclear transfer a public spectacle on a global scale. On the other hand, I wanted no part of mass media agendas that would fashion a monster of a minority religious community and its female scientist from one rib of the complex, living Raelian movement.

The story had disappeared even from the tabloids by early January 2003, about the time that Michael Guillen, a journalist who had organized scientific experts to authenticate the cloning experiment, abruptly broke off relations with Clonaid, suspecting a hoax. Yet this was too late for some Raelians – for example, the gentle music therapists who lived in the farmhouse adjacent to Rael's residence and were under invasion by paparazzi trampling their subsistence gardens and peeping through their ground floor windows. Email messages to me conveyed their distress. And since they had generously invited me to stay with them during an early phase of my fieldwork, I wondered if they somehow connected me, my notebooks, my tape recorder and camera, to the press invasion.

Dr Boisselier had in the meantime covered her raven-black hair with a fetching red wig and flown incognito to other continents to 'clone' more babies (including an Adam) who would later claim ever-diminishing minutes of fame and media space. Was she, then, 'the dove or maybe the raven' carrying a message of hope to humankind, along the lines of the fictional Nili? Would the world be more open now to new reproductive technologies, or less? Friendlier now to cyborgs and clones, or less friendly? Was she bringing science into closer step with the 'religious turn' (Weber, 2001) in modernity? Or was she imbuing all types of cloning and genetic engineering with the dark aura of cult practices?

'Oxygenation'

It is 2001, 55 A.H. (After Hiroshima), a year before Eve, and I am attending a gathering for the faithful and other interested persons in the farming and industrial outskirts of Valcourt, Quebec Province, on the residential grounds of the prophet. The occasion is the Annual North American Awakening Seminar where I am one of the 400 participating adults – a self-stated social scientist among committed or sympathetic Raelians, their guests, a couple of journalists from the *Sunday Times of London*, and (we later learn) at least one underground spy from the anti-cult establishment in Montreal. All of us mingle on the wooded grounds, which contain condominiums, a meeting hall, a modernist-style auditorium, office space, and the UFOland museum (then the world's largest ufological theme museum). There is an expansive lawn, a reed-lined pond, and a campsite with cabins and space for tents and trailers.

And now it is 9 a.m. and we have gathered for the signature morning ritual of Sensual Meditation. Most of us are wearing simple white poly-cotton robes, or *jalabas*, that could be purchased at the time of registration, as we settle into the plush theatre seats and tune our Walkman radios to channels where the French Canadian spoken here will be translated simultaneously into English, Japanese, Spanish, and Portuguese. The meditation content varies from day to day. Newcomers are informed that these exercises are 'nothing dangerous' and totally voluntary. We are told that the exercises will aid us in opening the machine of our body, opening the machine of our brain, and not to worry if we don't get quick results – we are seeking an efficient result. Now close our eyes… We are going to go inside the machine of our body. Air is entering our body and leaving it again. Deeply, deeply. Oxygenating ourselves consciously. We shouldn't be afraid of oxygenating ourselves. As we breathe deeply in, deeply out, our internal computer will increase its efficiency. The oxygen – 'compliments of the Raelian Religion' – allows us to clean our body of toxins. 'While you oxygenate yourself, listen to yourself breathing, your own rhythm is the rhythm of your genetic code … we open ourselves like a flower, let ourselves be penetrated by infinity.' 'Our cardiac rhythm slows down' and 'we feel the large black eyes in the sky that look down at us with so much love.' And now our voyage of harmony is about to begin… A wave of heat rising from our toes slowly upward to our brain, to the luminous spot, like a laser dot, between our eyes and above, like a spiral… Now we will begin to leave the room. We travel through the roof, we have a global vision of ourselves… And when we are ready, we will begin to move back again down toward the planet Earth, toward the blue ball, as if we are in a space shuttle. We see Africa, all the continents, as we go back through the clouds, gently. We try to spot the gardens, the little lake … the cabins… And now we re-enter the atmosphere of this great hall, all these people in white, in harmony. We are going to sculpt our brain in our way – to build a new life…

In the last and most intense meditation we are led in imagining our own deaths, leaving our bodies behind. As Rael did, we travel to another planet, are reborn there, and then re-enter Earth's atmosphere, being given our bearings when we hear that George W. Bush is unfortunately still the President of the US.

'That was fucking sinister, don't you think?' The London *Times* photojournalist approaches me as I drift out of the auditorium after the final meditation. 'I mean, Rael will revive you after you've died.' I look at him as if he'd come from another planet … and realize how pleasurable had been my abduction by the voice of 'technoscience spirituality'.[5]

'And the Spirit of Brigitte Soared Above Them'

It is another day, and some say a dissonant environment to the contemplative tones of meditation – the 'bordello' environment of the midweek open-air disco. The starry night is warm. We stand around, waiting for the music. Huge speakers on black metal stands create a ring, the prosthetics of a sophisticated sound system. Then they begin pounding out popular Quebecois, American, and World music. Eventually someone motions for me to join in the dance, and I do. And while dancing around I notice that Rael has arrived in his gleaming white Jeep Cherokee with his young wife Sophie. Brigitte Boisselier arrives at the same time. As always, Rael is dressed in tailored whites. Brigitte wears a little black party dress. This is the woman one anti-cloning leader of the religious Right has referred to as 'The Devil'. She and I have talked before, had critical exchanges on the subject of marketing designer babies, the racial, the ethical implications of this. But just at the moment we both want to dance. She waves and smiles and we end up dancing with each other. Then someone calls her away from the sidelines. I watch them talking. He leaves. She returns, leaning in above the loud music to tell me that the US Food and Drug Administration has just sealed her cloning laboratory.

Recall: it is summer 2001, a time when public, religious, bioethical, and Congressional interest in the cloning debate is front-page news. Recently, Boisselier has spoken by invitation to a US Congressional committee, articulating her faith-based commitment to human cloning as a matter of pro-choice reproductive rights, her commitment to helping couples who wish to produce a baby, for example gays and lesbians, or the terminally ill man who declares: 'I want to be reborn like a blank tape with the possibility to live only what I enjoyed.' She cites couples like the one who is currently sponsoring her laboratory research: their 10-month-old child died accidentally in a minor surgical operation. The parents were wishing to have him cloned so that the clone could one day stand before his surgeons and accuse them in court of his own wrongful death. These parents will later withdraw their support under intense media and government scrutiny. They will call Boisselier a 'publicity hog'. But tonight she is away from all this. She shrugs. And then she starts to dance again. I am amazed. I ask her, 'How can you dance at a moment like this?' And she responds, 'What can we do? We must dance.' And as her smile returns, the only word to describe her countenance is 'beatific'. I realize at this moment that I am dancing with a martyr. And indeed in the days to come, Brigitte Boisselier will announce to the Raelians and others at the morning assembly that she will take her lab offshore if need be, that she will not be stopped in her mission to clone a human being – to explore the frontiers of science. A member of her audience in Valcourt will later write:

> Did you see her move on the stage when she came to talk to us about the latest scientific news? She wasn't walking, she was floating, soaring … visibly from happiness! Who knows, in the future History books perhaps we will read something that goes like this: "And the spirit of Brigitte soared above them, and their eyes were opened and they knew they were like gods" … Brigitte states to whoever wants to listen, that she will fight until the end to see her projects come to light. (*Apocalypse*, 2001)

And who would doubt it? Expelled from France as a 'cultist', relinquishing legal custody of her children as a consequence, Brigitte Boisselier's is a lived political narrative of heroic self-sacrifice. Later, in a follow-up story to Baby Eve, an anonymous FDA official will tell the New York Times that the lab was an unsterile room in West Virginia where an unprepared graduate student worked with cow ovaries from a nearby slaughterhouse, using state-of-the-art equipment the sad couple of the 10 month old had financed (Kolata and Chang, 2003: A13). Raelians I queried were indifferent. Truth? Or a fiction of a hostile government? Or a complicit press?

The open-air disco is winding down. The revellers join together spontaneously in singing Rael's song to the Elohim, their arms raised to the stars. I am told, 'We gave them a sign – they must have seen it. They must know that we love them.' Which is also why Rael is raising money to build his Embassy on Earth. Conceived as a civil space with plush quarters for conducting diplomatic exchanges with the alien who is Us, the embassy is the ultimate faith-site of the Movement and optimistic countertext to Rael's vision of the apocalypse: the touchstone of what Harding and Stewart have termed 'optimistic apocalypticism' (2004). And you could see it at the time, on the Rael.org Web site, in posters on sale at the seminars, and prominently displayed as a model at UFOland alongside the crop circle design that inspired its architectural floorplan.

'Sweet People'

I depart the summer seminar in time to travel to the National Academy of Sciences Panel on Scientific and Medical Aspects of Human Cloning, where anti- and pro-reproductive cloning scientists, including Brigitte Boisselier, had been invited to engage one another before an NAS advisory panel. A bank of television cameras has trained its lenses on the stage. Reporters rove. But for many the action is offstage in the atrium where individual scientists, including a very agitated Ian Wilmut, famous for cloning Dolly, offer their viewpoints to journalists as audience members mingle over coffee and muffins, listening in. It is here that I spot Randolfe Wicker, founder of the Gay Cloning Rights organization and outspoken advocate for human cloning. A powerfully built bodyguard stations himself two steps away – Randolfe Wicker, he explains, sometimes doesn't know when to stop talking. I introduce myself to Mr. Wicker as an anthropologist studying the Raelian Religion. He is pleased to meet me.

'These are sweet people', he tells me. 'Personally, I like the Raelians. They were very nice to me when I spoke at one of their seminars. I got a standing ovation… And you know they are very pro-gay. But they turned the whole human cloning debate into a circus…' Randolfe Wicker had withdrawn his support for Brigitte Boisselier and Rael, whom he had once seen as kindred spirits, over the issue of objectifying women. There they were, young women parading themselves in front of the seminar audience, 'exhibiting themselves' as volunteers for human cloning or surrogates for extraterrestrials. But the rest was lovely; the seminars were wonderful. Outsiders 'were going there to their meeting thinking they were going to see people having sex on the grass… And then they saw they were normal people. There were couples that were going.'

Suddenly, our conversation is interrupted. Someone across the atrium is shouting 'It's immoral. This doctor [a presenter at the morning session] said in his own words that 70 per cent will be mutating. Maybe 99 per cent.' A badge identifies the protester as a Christian Against Cloning, a member of an organization opposed to embryonic stem-cell research. Wicker moves quickly to confront him and soon the two men are shouting back and forth so heatedly that members of the press rush to the scene, jostling for position. I find myself in the middle of the two and opposite a CNN reporter: she winks at me and smiles. We are together where the action is.

Randolfe Wicker: It is immoral to limit science, it's immoral to violate peoples' reproductive rights. That is what is immoral. They have the right to have their children; whatever happens it is their responsibility.

Christian: What will happen to their children?

RW: They'll turn out fine, or they'll miscarry or have to be terminated. I am sorry. It is the necessary price.

Christian: How do you know that child is mutated? How do you know when it's born? Do you recall what happened in Nazi Germany fifty years ago? Did you agree with what happened?

RW: I know very well. And you know what happened? Politics. Politics got involved. The State got involved, and decided who could breathe and how they could breathe. What was wrong fifty years ago was forceful sterilization. The same thing you do when you deny infertile couples access to human cloning technology. It is the same as forcefully sterilizing couples. And that is what you are trying to do…

Christian: Absolutely not. We are saying that science should not create monsters… It's a Pandora's box. Seventy per cent! Seventy per cent mutated children. Seventy per cent born deformed. We don't want humans to be used as laboratory rats. We saw this a generation ago in Nazi Germany.

RW: When the first human clone is born you will lose this debate. I'm fighting for reproductive rights and also for religious freedom rights of those people who feel that they have the right for their genes to survive until another lifetime.

'A Number'

In *A Number*, a play by Caryl Churchill that was first performed at the Royal Court Theatre in London on 23 September 2002 (National Science Year in the UK), the curtain opens on a man in his sixties and his grown son. The boy has discovered that he is a clone, created at his father's behest from a son he has lost, and further, that he, too, has been cloned, numerous times, by a rogue scientist who has long since died. The plotline unfolds as each of three cloned siblings confronts his father, and comes to terms with the idea of the others. As noted by the *New York Time*s theatre critic Ben Brantley (2004: B1), the men 'speak in shards'. 'Inarticulateness', he continues, 'has become the only fitting form of eloquence' in a world where people 'unthinkingly objectify others'. SALTER is the father; BERNARD (B2), the son.

B2: A number
Salter: you mean
B2: a number of them, of us, a considerable
Salter: say
B2: ten, twenty
Salter: didn't you ask?
B2: I got the impression
Salter: why didn't you ask?
B2: I didn't think of asking.
Salter: I can't think why not, it seems to me it would be the first thing you'd want to know, how far have things gone, how many of these things are there?
B2: Good, so if it ever happens to you
Salter: no you're right
B2: no it was stupid, it was shock, I'd known a week before I went to the hospital but it was still
Salter: it is, I am, the shocking thing is that there *are* these, not how many but at all
B2: even one
Salter: exactly, even one, a twin would be a shock
B2: a twin would be a surprise but a number
Salter: a number any number is a shock.
B2: You said things, these things
Salter: I said?
B2: you called them things. I think we'll find they're people.
Salter: Yes of course they are, they are of course.

> *B2*: Because I'm one.
> *Salter*: No.
> *B2*: Yes. Why not? Yes.
> *Salter*: Because they're copies.
> *B2*: copies? they're not
> *Salter*: copies of you which some mad scientist has illegally
> *B2*: how do you know that?
> *Salter*: I don't but
> *B2*: what if someone else is the one, the first one, the real one and I'm
> (Churchill, 2002: 10–11)

But the last word in the play is Michael's, the youngest clone's. Unlike B1 or B2 his name is fully written in the script, he has a family, a social context, a social future. He adoringly describes his wife's 'slightly odd ears ... slightly pointy on top, like a Disney elf or little animal ears' – like a loveable transgenic, not apart from other creatures, other 'things' like the dogs that Michael likes; perhaps not apart from the alien-human hybrids of ET culture, the 'grays', whose distinguishing feature is their pointy ears.

> *Salter*: And you're happy you say, are you? You like your life?
> *Michael*: I do yes, sorry. (Churchill, 2002: 64)

'The Girls From Greenwich'

In the first season of the cult television series *The X-Files,* an episode titled 'Eve' tells the story of a female scientist who, after being abducted by aliens, begins secretly to clone herself. Her illegal activities are discovered by The Government, which succeeds in locking up some of the Eves for study. But others escape, and their eggs make their way to *in vitro* fertilization clinics where they are used to generate a third generation of (alien hybrid) Eves. These hybrids, however, in addition to being uncommonly strong with mega-genius intellects like their Creator, demonstrate sociopathic behaviour – specifically, they take to killing their fathers. The 'bad seed' of alien science is propagating a matriline – and potentially a matriarchy – of monsters. And like the Hitler clones of the well-known replication film *The Boys From Brazil* (an earlier title for this *X-Files* episode was 'The Girls From Greenwich'), these Eves live undetected in an iconically normative pocket of the country, waiting to strike; their traces voids where males once were (Lavery, Hague, Cartwright, 1986).

It is a matter of speculation whether the Eve of *The X-Files* took her name from the Eve of Christian mythology; or whether the Raelian Eve was meant to retell the former as a counter-narrative of feminist hope, or the latter as a critique-by-appropriation of Christian dogma. The point here is that, as a collective, the cultic Eves of speculative science engender an alternative map for the future of the 'species concept', and for human social networks oriented to that futurology. Faith sites of a new vision of kinship, they inhabit neither the mythical space of recursive time, nor historical chronologies, but instead a mediatized 'event-time' (to take a concept from Deleuze, 1989): a temporal mode for drawing the social, and specifically social memory, into contingent futures of relationship. As such they are compelled to abandon their contexts, their own seriality and lineage horizon. For they have engaged us, as readers or audiences or ethnographers, sceptics or believers, in the serious play of recombinant community-making.

'Trickster'

In November 2004 I am invited to give a version of this paper in Heidelberg, Germany, at the European Molecular Biology Laboratory (EMBL) fifth Annual Conference on Science and Society. The invitation is extended by the lone anthropologist in this highly respected organization, Halldor Stefansson.

The theme of the Heidelberg conference is 'Time and Aging – Mechanisms and Meanings'. The audience is composed mainly of biology students, teachers, and research scientists, as well as members of an educated public, most from northern Europe and the UK. But Halldor Stefansson, a self-described 'trickster' along the lines of the 'somewhat monstrous' Norse figure of Loki, is accustomed to the challenges of mediating the worlds of 'hard' and literary social science.

His diplomatic skills will be tested on this occasion. Already during my talk, I can hear a voice from the front row (could it be the EMBL Director's?) exclaim, 'She's a Raelian!' And while many in the audience are delighted by the Raelian ethnography, others are deeply disturbed. One keynote speaker, a dignitary of the Church of Scotland, is particularly shaken by the Raelian vision of the god-function. A doctoral student confronts me after the talk, furious that I am using this forum to 'proselytize' for the Raelians – a point of confusion that has never before emerged. So I email my host, withdrawing the paper I had promised for publication and saving him more grief. To which he responded with his customary tact and elegance:

> Indeed, there can be surprisingly abrupt limits to the common ground where professionals of different disciplines can engage fruitfully in a dialogue. The mostly sympathetic depiction of the Raelian Other you brought up proved too much for the sensibility of many of the scientists in the audience in Heidelberg. I guess you could say that you presented them with a mirror where they saw their own image seriously distorted. The choice they were presented with was between identifying with that 'monster' in there, or rejecting you. I guess the outcome was easily predictable.

I appreciated this analysis. Both Raelians and ethnographers find our natural habitat in the spaces between known and unknown, visible and invisible cultural orders. The questions we ask about what it means to be human have drawn us to these spaces beyond the borders of familiar environments and subjectivities. And as Mary Douglas (1978) famously theorized for 'cultural orders' ambiguously threatening interstices, and Victor Turner (1967) for the 'betwixt and between' realms of cultural performances, we turn to such liminal zones in our search for monsters. Indeed, in a sense we are disciplined to seek them there, as figures of anomaly, abjection, excess, alarming ambiguity.

Yet structural models that presuppose a bounded society or culture, fixed normative orders, a coherent value system, and demarcated border zones for setting categories of the human apart, are not well-suited to the galaxies of discourse (Battaglia, 2006) that converge, diverge, and reconstellate under the influence of entities such as Eve; cannot in and of themselves represent the anxieties of what Zygmunt Bauman (1991) has termed 'fluid modernity'. Neither are they sufficient to the aberrations we might locate within the fluid Eve communities I have drawn here – communities composed of found subjects and found objects, of human and non-human agents (Latour, 1999: 284), in relational flux.

In this light, we might expect the mass media, and especially new media, to be predisposed to participate in this story of emergent life (cf. Lister et al., 2003: 304), if only because digital speed further blurs defining fields before they have fully formed, while networking practices make it difficult to know where personal or collective agency begins and ends; difficult to access (much less evaluate) the programmes streaming through our living spaces, or to escape informatic black holes created in their wake. The Raelian programme manifested as Baby Eve puts a finer point to this condition. Clearly it was not only the sin of media speed that caused concern for my audience in Heidelberg, some of whom had actually participated in the 'race' to complete the Human Genome Project and to copyright its codes. Rather, it was as if I had conspired with the Raelians to send a lethal abyss-producing programme into the science community, maligning its connective tissue, in very real economic and political terms threatening its entity livelihood, its circulatory pathways of information, its exchange networks.

In short, whether we focus our inquiry on contact zones or on aberrations of flow, respectively, each model displaces and denies a greater anxiety – which is the anxiety of the abyss: a dimensionless chasm that defies spanning or the reach of our technologies of knowledge, pre-defeating connection. There is no future in the abyss, only abnegation.

The irony is that Baby Eve, trickster entity of the breaking moment of news, helped to establish 'the Raelians' as a brand name of marginal science practice. When, for example, Advanced Cell Technology announced the extension of cloning experiments to cow-human embryos, religious fundamentalists could now be imagined as thinking, 'My God ... these people are going to make chimeric creatures – mixing cows and humans'; the CEO of another firm who stated this adding, 'It's not in the same category as the Raelians, because there are certainly legitimate scientists at ACT trying to do this work, okay? But from the perspective of the regulatory bodies, they are in the same spaceship' (Anonymous, quoted in Dunn, 2002: 40).

Two months after leaving Heidelberg, I would read in the *New York Times* that Halldor Stefansson's brother Kari, a scientist at DeCode Genetics in Reykjavik, Iceland, who was also an invited speaker at the EMBL conference (though better received to say the least), had made an amazing discovery. The discovery was of a region of the human genome – a vast section of DNA – that had 'flipped' or become inverted in some Icelanders, while running in the standard direction in others. Because the flipped (alien) and standard versions do not match, the two cannot exchange genetic elements during recombination – that is, during the shuffling of genes that gives rise to genetic novelty in successive generations. Kari Stefansson and his team hypothesized that the flipped region was carried by archaic proto-humans. At some point it had come adrift from its parent chromosome and only later re-entered the modern human lineage in some episode of interbreeding with modern humans, 'knitted back together the wrong way round'. The archaic, flipped (alien) version increases the rate of reshuffling; is more 'creativity-prone' in this sense. It also 'seems to confer on people the ability to live to extreme old age', according to Dr Stefansson (Wade, 2005: A6).

How Raelian scientists would interpret this startling moment in the human creation narrative of DeCode I don't yet know. But I can guess. Here is a statement of Rael's from a raelscience.com email message that was circulated shortly after I arrived home from Heidelberg:

> The Raelian Movement would like to underscore that The Theory of Intelligent Design does not lead to a supernatural designer but to an extraterrestrial human civilization designer, which is in line with naturalist philosophy. One day, maybe sooner than you might imagine possible, we will go to other planets and scientifically engineer life on them. Then, probably, these worlds will create deistic religions from the interactions we will have with them. *In short, what we need to be focusing on is not 'survival of the fittest', as microevolution sees adaptation, but 'arrival of the fittest'* [emphasis mine].

'The Rael World'

> 'The longing for immortality of many cultures is not for more of the same but for a world different from this one.'
>
> Leon Kass on National Public Radio, 15 April 2002

When the Eve story finally died it was Rael himself who brought the movement back to media life – in televised interviews, and in the 'Men's Fashion' supplement to the *New York Times Magazine.* Elegantly seated in his formal white uniform that resembles the alien messenger Klaatu's in the Cold War film classic *The Day The Earth Stood Still* (1951), he had been voted 'Our Favorite Prophet' by the magazine. The same magazine would later put him on their celebrity chef list, and quote him on his recipe for duck à l'orange. The New York media, it seems, was getting with the Raelian programme, at least to the extent of re-engineering its image.

Meanwhile, mainstream media were elsewhere covering the disrobing-for-peace ritual of Raelian 'angels' at rallies protesting the war in Iraq in Washington, DC and Los Angeles.[6] Then, after a long lapse in visibility, *Playboy*'s November 2004 issue ran 'The Rael World', a lavish pictorial feature that included six pages of nude Raelian beauties. One could read in the Table of Contents: 'The group of UFO believers wants to clone its most beautiful members. We believe!' Raelians, it seemed, were returning to the notion that human beings were 'the Elohim's supreme artworks'. Inside we could read that Brigitte Boisselier's daughter, Marina, who posed in front of colourful beakers, was producing a documentary film on embryonic stem cell research. Marina 'believes the secret of immortality can be unlocked through stem cell research and cloning.' She is quoted as saying 'The Raelian philosophy doesn't want everybody to be the same but encourages you to get in touch with who you always wanted to be and reprogram yourself to your own taste... Not only do we respect differences but love and encourage them.' A photo gallery shows two Raelian women kissing with a beaker in the foreground; Raelian 'angels' marching for International Women's Day in Los Angeles, and the RMX2020 embryonic cloning machine developed by Clonaid: the machine that was used to produce Baby Eve.

Had the 'mad scientist' seduced the media once again, stealthily drawing attention away from the invisible Eve to Marina, her very visible flesh-and-blood offspring? Or drawing attention away from reproductive cloning to the next generation of (therapeutic) cloning discourse? Or were media apparatuses, and the 'heat' of celebrity, doing the work of creation?

Of Cults and Monsters: The Disruptive Molecule

This is what I know about prions. A prion is a protein molecule that has no genetic material of its own; no agenda to replicate. It is produced mainly in the brain but also in the pancreas. And like all protein, it exists in a three-dimensional folded shape. This shape may be 'normal' or it may be 'disruptive'. Either way, it can form a *template* for other molecules (as Stanley Preusiner demonstrated in his pioneering research). For example, a disruptive prion has the capacity to attract others to it, convex bulges conforming to concave spaces and vice versa, as they heat up and form sticky clusters, resembling a lint ball. Their chemistry is weak, but effective. And the *scale of replication* does not matter: the part – the prion – can mortally disrupt the whole of the organism. As the long chains gather and cohere, the cells in the tissue-context break apart, and the tissue breaks down. Most famously, this prionistic activity is the cause of Mad Cow or Creutzfeldt-Jakob Disease (CJD), and of the spongiform brains (and we now know, other parts) of its victims. But too, dangerous prions can as well be inherited – 'witch prions' evident at birth or else dormant as a destructive potential.

In this association the prion is an archetype of 'monstrosity within' the body. And because the discourse of proteins is relatively neutral, at least as compared to genomics, the prion has tremendous trope-ic potential as an image of and for the dangers lurking within the body politic; within the body of the scientific community; within minority religions. Getting to the source of a prionistic infection is like searching for accountability within a complex system (human, political, economic) experiencing a highly covert and hostile take-over. Until the disease's final stages, it is difficult to tell, for example, the healthy cow from the cow that could be the death of us all. The animal is not a monster, although its body chemistry might express a monstrous proteinaceous programme – a malevolent proselytizing agent.

The words 'cult' and 'sect' evoke this Mad Cow quality. Under cult influence, we fear that that we or those we love will become 'spacey' and unpredictable. The cult's programme locks us into its parameters, prescribing and then opening yawning gaps in knowledge exchange. Of course not all programmes exert the deadly influence of a malformed and malforming prion (and not all religious movements, which tend to be characterized by their incoherence and seem to be making themselves up as they go along, are programmatic). For that matter, even disruptive prions may have positive functions we do not yet understand. But under the over-determined influence of the labels 'cult' and 'sect', how naturally the map is taken for the thing.

Court TV's Web site described Brigitte Boisselier as 'a leader of a sect'. As a follow-up, the station opened its phone lines to callers who had questions for Lori Andrews, author of *The Clone Age,* and documented their exchanges. One caller asked her opinion of engineering a 'perfect child' (Clonaid's eugenic marketing had opened Boisselier to charges of Internet fraud). Andrews responded that 'a March of Dimes poll showed that forty three percent of parents say they would genetically engineer their kid to give them better physical abilities and forty two percent would enhance their kids' mental abilities. Twelve percent of parents say they would abort a foetus with a gene for obesity.'

Overall, you will probably have noticed that this essay is itself something of a teratological artefact. Experienced as an academic talk, which is how I first conceived of the paper, you would have, in addition, heard parts of Rael's song to the Elohim. You would have seen projected images of Rael playing an acoustic guitar, of the phenomenally mediagenic Brigitte Boisselier on the front page of tabloid newspapers, of T-shirted Raelian youths laughing together in a circle on the lawn, of folks sunbathing in the Garden of the Prophet. I like to imagine that my academic talk elicited a sense of the melting boundaries of anthropology's classic relationship to magic, science, and religion; and perhaps in its function as embodied description, invited an effect of *productive anxiety* for our diverse projects of writing culture, and more generally for the stability of the notion of 'context' so revered by the discipline of anthropology.

On another level, I have attempted to bring an ethnographic presence to the question of where we find our monsters in the New World Order of our networked, media-enhanced modernity; meanwhile claiming a value for methodological openness. Thus, while I began my anthropological itinerary at the point of doing fieldwork with Raelians, I found it not just constraining but inaccurate to omit engagements with texts and images that in some respect asked to be related, or analogically juxtaposed.[7] Put another way, the essay quite literally *re-plays* the creative action of the emergent relations it writes.[8] Giving itself to dialogue with agents outside of the moment of 'being there' alongside Raelian subjects, it embraces a methodological teratology. And along the way it pays homage to the 'gaps not yet the abyss'[9] without which our leaps of faith in the social would be unthinkable.

Notes

1. This chapter is for Donna Haraway. Portions have been taken from my essays in *ET Culture: Anthropology in Outerspaces* (2006), with the gracious permission of Duke University Press. I wish to thank Mount Holyoke College for funding the research and Penny Harvey, whose invitation to address the Association of Social Anthropologists meetings in Manchester, UK, led me to experience the possibilities of performing ethnographic description. The extract from Caryl Churchill's play, *A Number*, is cited by kind permission of Nick Hern Books Ltd.
2. I borrow this phrase from the title of the last book of John Mack, whose efforts to legitimize his psychiatric patients' abduction narratives and experiences opened the door to de-exoticizing discourses of the extraterrestrial.

3. On Raelians, see Susan Palmer's extensive sociological research and her *Aliens Adored* (2004). Chryssides (2003) has also published on Raelians in England.
4. For more on this strategy, see Annelise Riles (2005) on 'legal fictions' in her paper of that title.
5. I define technoscience spirituality as 'hard-faith in science and technology, future' (see Battaglia, 2006).
6. Raelian planners knew only that the ritual was West African. It may be that the model is a turn-of-the-century practice recorded by a colonial officer in Ghana, where women shamed men into stopping their plans for war, by stripping down in public space and bragging like men about their 'war exploits' (Delafosse, 1913). I am grateful to Victoria Ebin for the reference.
7. Cf. Marilyn Strathern's equation of anthropological method with fieldwork (2005), which usefully insists that the discipline should not forget the value of context in an age of globalization models. For an important interrogation of anthropology's essentialized value of context, see Roy Dilley (1999).
8. Wolfgang Iser's (1989) early work towards a literary anthropology in which the importance of performativity and play is central, is a move in this direction.
9. This great notion is Giorgio Agamben's.

Bibliography

Agamben, G. (1999), *Remnants of Auschwitz: The Witness and the Archive,* New York: Zone.

Battaglia, D. (ed.), (2006), *E.T. Culture: Anthropology in Outerspaces*, Durham: Duke University Press.

Bauman, Z. (2001), *Community: Seeking Safety in an Insecure World,* Cambridge: Polity Press.

Brantley, B. (2004), 'My Three Sons', *New York Times*, 8 December: B1, B7.

Chryssides, G. (2003), 'Scientific Creationism: A Study of the Raelian Church', in C. Partridge (ed.), *UFO Religions*, London: Routledge.

Churchill, C. (2002), *A Number,* London: Nick Hern.

Delafosse, M. (1913), 'Coutumes observes par les femmes en temps de guerre chez les Agni', *Revue d'Ethnographies et de Sociologie,* 4: 7.

Deleuze, G. (1989), *Cinema 2: The Time-Image,* trans. H. Tomlinson and R. Galleta, Minneapolis, MN: University of Minnesota Press.

Dilley, R. (ed.) (1999), *The Problem of Context,* New York: Berghahn Books.

Douglas, M. (1978 [1966]), *Purity and Danger: An Analysis of Concepts of Pollution and Taboo,* London: Routledge.

Dunn, K. (2002), 'Cloning Trevor', *Atlantic Monthly,* 289: 31–52.

Haraway, D. (1995), 'Cyborgs and Symbionts: Living Together in the New World Order' in C. Hables Gray (ed.), *The Cyborg Handbook*, New York: Routledge.

—— (1997), *Modest_Witness@Second_Millenium.FemaleMan©_Meets_OncoMouseTM,* New York: Routledge.

Harding, S. and Stewart, K. (2003), 'Anxieties of Influence: Conspiracy Theory and Therapeutic Culture in Millennial America' in H. G. West and T. Sanders (eds), *Transparency and Conspiracy,* Durham: Duke University Press.

Iser, W. (1989), *Prospecting: From Reader Response to Literary Anthropology*, Baltimore, MD: Johns Hopkins University Press.

Jacoby, S. (2005), 'Caught Between Church and State', *New York Times*, 19 January, A27.

Kolata, G. and Chang, K. (2003), 'For Clonaid, a Trail of Unproven Claims', *New York Times,* 1 January, A13.

Latour, B. (1999), 'One More Turn after the Social Turn...', in M. Biagioli (ed.), *The Science Studies Reader*, New York: Routledge.

Lavery, D., Hague, A. and Cartwright, M. (eds) (1986), *'Deny All Knowledge': Reading the X-Files,* Syracuse, NY: Syracuse University Press.

Lister, M., Kelly, K., Dovey, J., Giddings, S. and Grant, I. (eds) (2003), *New Media: A Critical Introduction*, London: Routledge.

Mack, J. (1999), *Passport to the Cosmos: Human Transformation and Alien Encounters,* New York: Crown Publishers.

Palmer, S. (2004), *Aliens Adored: Rael's UFO Religion*, New Brunswick: Rutgers.

Piercy, M. (1991), *He, She, and It*, New York: Knopf.

Rael (2005 [1998]), *Intelligent Design – The Message From the Designers (e-book).* Available at: http://www.rael.org.

Riles, A. (2005), 'Legal Fictions', Paper presented at the American Anthropological Association meetings, Washington, DC, November.

Strathern, M. (2005) 'Useful Knowledge', Isaiah Berlin Lecture, 2 December, British Academy, London.

Turner, V. (1967), *The Forest of Symbols: Aspects of Ndembu Ritual*, Ithaca, NY: Cornell University Press.

Wade, N. (2005), 'Genetic Find Stirs Debate on Race-Based Medicine', *New York Times*, 11 November, A5.

Weber, S. (2001), 'Religion, Repetition, Media', in H. de Vries and S. Weber (eds), *Religion and Media*, Stanford: Stanford University Press.

–9–

Echolocation in Bolivip

Tony Crook

> Relativity has always been vital to anthropology, which underwent many of its formative crises and transformations contemporaneously with the development of relativity in physics.
>
> Roy Wagner, *Invention of Culture* 1975: 152

> A Nietzschean idea that interpretations and points of view are constitutive of objective facts is worth reconsidering for any negotiations between culture and science.
>
> Heonik Kwon, panel abstract 'Perspectivism from an Anthropological Perspective', *ASA Decennial,* 2003

Arguments Among Myths

Echoes depend on location. Perhaps this suggestion that an echo – what the world returns to one – depends on location sounds too obvious for words. But in February 2003, the *New Scientist* magazine carried a worrying cover story: 'Was Einstein Only Half Right?' Coming shortly before the centenary year of Einstein's famous series of papers, a revolutionary theory – called 'doubly special relativity' – dared to suggest that the long-held reliance on an invariant speed of light was unable to account for some recent observations, and may possibly prove to be a stumbling block to reconciling the fields of quantum theory and gravity in physics. Doubly special relativity (DSR) questioned the uniform speed of light by suggesting that greater energy at the beginning of the universe may have caused light to travel faster then – in effect, the measurements over recent centuries reflect the same moment in a slowdown. Measured against another time or phase of activity in the universe's creation, the speed of light as currently measured might yet turn out to be relative to our own position. Given the particular importance of the speed of light in the equation that unified energy and matter (the c in $E=mc^2$), the physicist Amelino-Camelia's comment that 'Einstein may have had only half the story' was all the more resonant. And because light has, at least since the time of Plato's famous cave, provided a long tradition in epistemology with a ready metaphor for knowledge, it seems apposite to think through ideas about the position of an observer, and the variation and variability of knowledge. The senses, too, vary in the subject-object divisions they afford – epistemological metaphors based on sight might be augmented by considering those based on sound. I do so by resounding ideas of relativity in physics and anthropology through Melanesian and Amazonian materials, catching what returns from these 'arguments among myths' that 'speak to one another ... protest, alter, and enlarge upon the myths of the other' (Gillison, 1993: 3), and listening for resonances defining the subject's position by what Wagner (2001) calls 'echolocation':

> [B]ats locate themselves, navigate, and find their food by bouncing sound off of echo-limits, effectively transforming their negotiable world into an imaginary crepuscular cave... Human beings [however] locate their subject, its negative spaces or contingencies, by resonating against the limits of language... One listens for intent in one's own speech as well as in that of others, and one calls that intent 'meaning'. (Wagner, 2001: 136–7, 138)

'Echolocation' then, acknowledges that variability comes of and returns to a subject who acquires definition in these relations: in short, it matters very much who is occupying the position.

In rethinking the physics of the natural, theoretical and social worlds around him, Einstein developed various thought-experiments as a means of taking soundings from each in terms of the other through the mediation of his own position in the scheme. For example, the changing sound of a moving train when approaching or going away from an observer led to thought-experiments asking about light: whether, when riding on the front of a beam of light, Einstein would be able to see himself in a mirror. In this scheme, observers' knowledge depends solely on his position – their internal disposition or composition is of no consequence here. Put bluntly, it makes no odds who it is occupying the position. As with the impression made when hearing and seeing co-ordinated town clocks mark the hour, the point here for Einstein was to take the observer's position *out* of the equation. The echoes of sound and light may have proven useful in describing relative locations but the real question remained. In his *The ABC of Relativity* (1925), Russell derided 'a certain type of superior person' who imagines that 'the new theory proves *everything* in the physical world to be relative, whereas, on the contrary, it is wholly concerned to exclude what is relative and arrive at a statement of physical laws that shall in no way depend upon the circumstances of the observer' (Russell, 1925: 14, original emphasis).

In promoting the paradigms of 'one nature, many cultures' and 'one observer, many interpretations' – locating the contingencies of culture and observer in the equation – generations of anthropologists have been arguing quite the reverse. An observer's self-knowledge through relations with others and with the world has been important in shaping understandings of anthropological knowledge – acknowledging the presence and involvement of the ethnographer made a virtue of this rather obvious necessity. However, equating personal and textual involvements in relations and literary culture with the diminishment of objectivity makes plain the continued entrapment in the associations between relativity and subjectivity, and rationality and objectivity, as though these were necessarily the limited options available. For example, Bateson's admissions in *Naven* (1958 [1936]) about the shortcomings of his study lead Radcliffe-Brown to describe the work as an 'intellectual biography' and another reviewer to worry that 'Bateson's theory ... includes too many personal elements to be called, without qualification, scientific' (cited in Lispet, 1980: 146). But must an involvement in one's interior life and relations with others necessarily implicate *a point of view*? Imagining a position in relations as compromising an objective perspective tells of a reflex to divine the possibility of an outside observer looking into a landscape. For Bateson's reviewers the bystander seems to have been science in the strict sense of the term. Perhaps those limited options, and indeed the model of relativity in physics, make unhelpful metaphors for what anthropologists mean to say?

Relativity is a marvellous metaphor for individuality. The multiplicity of points of view makes for a scarcity if they are to be individualized, and attributes value and power to knowledge that is scarce. One mark of an individual is that individuals have their own point of view and, rather than diluting individuality, any revelations about their interior dialogues or relations with others bolsters the relativity of their individual position when considered against some imaginary objective benchmark. And whereas solipsism bares the solitude of an individual, relations with others bears out an individual's individuality. With this double rationale to hand, either being alone or being with others becomes a path to refract or define individuality in particular ways. Curiously though, when a point of view becomes a collective position shared with others, their individuality disappears – their internal disposition or composition is no longer of any consequence. Strangely, it makes no odds who it is occupying the position. Might we add 'collective' and 'individual' to the list of mutually implicating limited options?

One origin for the 'Perspectivism from an Anthropological Perspective' panel was to consider Viveiros de Castro's example of Amazonian perspectivism (1996) in relation to the philosophical development of perspectivism and other responses to Einstein's theories of relativity in the early twentieth-century. In Viveiros de Castro's scheme, the variation that comes of different beings, such as jaguars or humans, occupying the subject-position of observer and seeing the 'same' thing, such as blood or beer, does not depend on the internal disposition or composition of the observer. What matters here is that perception depends upon, and confirms, whether the observer-being is animal or human. Whereas age, sex, gender, aptitude and respectfulness figure importantly in Melanesia, clearly, they are of no consequence here. Interestingly, it makes no odds who it is occupying the position. For Ortega y Gasset the new theories of relativity had double significance: the shift from Newton's description of absolute space as '*sensorium Dei,* the visual organ of God', made plain a spiritual ambition to attain this 'divine perspective' on reality independent of the position of an observer, whilst Einstein's 'marvellous proof of the harmonious multiplicity of all possible points of view' (Ortega y Gasset, 1931: 143) constituted 'one more step on the road to subjectivism' (Ortega y Gasset, 1931: 141). Of course, this double move – dependence and non-dependence on the observer – echoes the persistent dual narratives of objectivity and subjectivity ascribed to the implications of Einstein's work, and which obtain their form through each other (see Barnes and Bloor, 1982). The stimulus for the panel then, was Ortega y Gasset's portrayal of a 'persistent error that has hitherto been made [: the] supposition that reality possesses in itself, independently of the point of view from which it is observed, a physiognomy of its own' (Ortega y Gasset, 1931: 91).

Now, one reason why the *New Scientist* article caught my attention is that junior male cultists in Bolivip live in hope of having more than half the story (Crook, 1999). My interest here is drawn to the attention that men, especially, give to locating their knowledge through others – and because 'knowledge' here is conceived as a bodily water that grows the skin, and because it is given over by other men in exchanges of care, there is an acknowledgement that a person's bodily capacities are located by others. Neither their person nor their knowledge is only their own – indeed, these exchanges are based on revealing inner capacities and result in two men sharing 'one skin'. As we will see, each man constitutes only half a point of view, so to speak, and expects the other to complete the story into 'one'. The analysis of Bolivip conceptions of knowledge as bodily substances suggests that variations in knowledge are better understood as revelations of a person's internal gendered composition – what I will describe as their 'echolocation' in the meanings, world-view and bodily resources of those who began their ongoing contributions in conception. The awkward maths of these materials appear to muddle relativity and rationality, subjectivity and objectivity, collective and individual and perhaps offer some hope of finding alternative options.

Angkaiyakmin are a Faiwol-speaking Min group of around a thousand people, originally from Telefomin and based in Bolivip village, and now dispersed throughout six villages in Western Province, and in urban centres across Papua New Guinea. The Min or Mountain Ok are perhaps best known by Barth's classic work on secrecy stemming from his Baktaman study (see Barth, 1975, 1987, 2002). The considerable problems that anthropology has encountered in analysing the Min make it a particularly revealing place in which to see the discipline's ideas of knowledge at work (Crook in press). Indeed, Herdt (1990) was able to illustrate both his 'pessimistic' and 'ontological' models of secrecy with examples from the area (Barth and Bercovitch, 1989, respectively).

The assumption that the power of knowledge is a virtue of scarcity is a central motif of Barth's monograph *Ritual and Knowledge among the Baktaman*: 'the value of information seemed to be regarded as inversely proportional to how many share it' (Barth, 1975: 217). Although Barth (1987) located the nexus for cultural variation amongst the Min in the melting-pot of a senior initiator's mind during extended lapses between ritual performances, he seems to have allowed no such creativity and variation amongst the initiates who all seem to share in the limited exposures and cultural transmissions during the rituals. Scarcity and variations lay only between the grades. The analytic lends a particular character to knowledge such that bounded units of knowledge are possessed by discrete domains of men, and carries a set of corollaries that enabled Barth to read levels of truth into a sociology of hierarchical initiation grades, and to locate power simply by tracking the distribution of secret knowledge revealed solely during infrequent rituals exclusively amongst men. Barth's analysis has grades of initiates in possession of the same level of knowledge – as if they were a collective-individual and all of one mind. In this scheme, an observer's knowledge depends solely on his position in the initiation grades their internal disposition or composition is of no consequence here. Put bluntly, it makes no odds who it is occupying the position.

Holography Amongst Halves

Not least because this was the position of my own anthropological apprenticeship, my discussion here attends to the circumstances of the observer in the person of a junior male cultist, and to his position and location in an inter-generational body of knowledge and relations. In doing so we might ask in what he is positioned and in whom he is located? The path ahead then, involves examining the 'halves' brought together in making and growing skin in knowledge and conception transactions.

For male initiates in Bolivip, growing knowledge is figured as moving down a tree from confusing movement in the crown to resolved stillness at the base from where senior cultists describe how the earlier branching effects of cult talk about *awem* ('important knowledge') reduce until all the fragments of *weng fakong* ('broken talk') come together towards *leip makop* ('one path') and *sung makop* ('one story'). This ideology of 'one path' supports many branches, enables ample creativity and affords the encompassment of infinite variation. Now, the ease of interpreting these variations and knowledge practices in terms of philosophical concerns with 'justified true belief', or anthropological concerns with instrumentality become problematic when we consider the descriptions given in Bolivip of 'knowledge' in terms of contributions of bodily substances and resources which are a complementary reversal of men's and women's differing accounts of conception theories. The passage figured as the descent of a tree involves the receipt of bodily resources from an older generation of initiates during a sequence of male initiation rituals, and through transactions with older paternal and maternal male kin in 'exchanges of skin' (care for knowledge). These 'showings' (*ban*) (known in the literature as male initiation rituals, see Barth, 1975, Poole, 1982, Gardner, 1983), involves revelations of sacra, names, stories and moral advice which impart *awem* ('important knowledge') through vision and language that becomes held in the skin (*kál*) in the junior man (see Eggertson, 2003). It seems to be a deliberate strategy amongst initiators to make the younger initiates dizzy: to turn their minds with words, revelations and contradictions, and to turn them again until, in their very dizziness, they come to see straight what could not be put into words (Jorgensen, 1990). These visual and audial revelations of 'knowledge' (*kal*) are conceived generally as 'advice' (*sawa*) and are said to carry and impart a bodily water (*wok*) that is evident in the shine of skin and in a shiny membrane (called *lamlam*) inside the thigh, where junior men are told to hide and keep their important knowledge.

The expectation of there always being another half to knowing in Bolivip comes of 'clear' aspects (*fitap*) having a 'hidden' (*ati*) base, and is often fulfilled by the forms of revelation. Expositions often involve juxtaposed fragments, which invite the listener to make the relation for themselves – junior cultists complain of being tricked when told 'only half' (*mari*) and describe the work of gathering and straightening little pieces of talk together (*kim kurukuru taretare*). For example, initiates are taken aside during ritual and have what they have been shown explained to them by maternal and paternal kin and afterwards know to ask again and again until they have things 'straight' (*turon*). An important source of verification in the work of 'straightening' derives from junior men checking their father's advice with a mother's brother. Even so, senior cultists describe how it was only when they took their turn to show an initiation that they had themselves been shown – seeing both halves together, so to speak – that they came to understand more of what appeared hidden at the time. These positions are also marked by gender in which 'males' reveal to 'females': when junior cultists first leave the house of their mothers they also leave *wanang am alin* ('those of the women's house') to join *kinim am alin* ('those of the men's house'), but in doing so they find that they are, relative to those more senior to them, still *wanang am alin*. They are made aware that divisions outside the men's house have internal counterparts. Although the ideology of 'one story' enables a sense of progression that comes of knowing that there is always more to know, men remain agnostic about conceptions of 'clear' and 'hidden' halves that can be combined to complete each other. They are made aware of their own part in a complex part-whole relationship.

The relative positions of junior and senior cultists are explained as positions in a tree: an image reiterated by juniors who describe asking questions of an old man only to find him unmoved – without any movement of face or body that might give something away, and who describe their own experience of learning *awem* as filled by confusion as a story seems to be moving around, branching off into many strands. Old men describe how juniors always seem to be talking of the same thing, always asking the same questions, whereas junior men say that old men do not think first before talking, and keep the important parts hidden. One young man recalled that whenever he asked a question, his father just gave him another example: 'Giving these old men questions only makes things difficult; they might be trying to help by bringing another example in but they often turn their words around and this just makes things branch more, and then you have to turn their words around again. We call this *weng fakong*, like breaking talk into pieces; it's the same as *iman fakong* when you break a taro in two and give the other half to someone.' Another man described the experience to me as *kutarkutar*:

> When a story does not go straight or the teller does not use straight-talk [*weng turon*], when the teller goes off and starts 'breaking a path through the forest' [*saak leip, saak weng* – unconnected path, talk], when the pieces or examples seem to be jumping about, this is like *kutal* the Black-tailed Giant Rat who is the most agile and hardest one to catch in the forest even with dogs, and might just jump from a tree, dart away and disappear.

He advised me that I should 'gather up all these pieces of stories and examples and put them all together as one story – *kim kurukuru taretare*.' 'You are still in the crown of the tree where all the fruit-water and flower-water hungry birds [*wok awon*] move around and sing at once, but maybe we can help you learn to see how all these pieces might be gathered up and come together into one sentence.' The image of revelations of *awem* as positions on a tree, then, suggests that for juniors in the crown, words from seniors at the base appear to branch into multiple possibilities; whereas for those at the base, juniors only demonstrate that they do not know by going on about the same thing.

Although the regionally important creator ancestress, Afek, failed to reach Bolivip, the paths of ancestral precedent (*aulal kukup*) are held to follow those created by her. Each stage in the sequence of *ban*, and each small part of important knowledge (advice, revelations, myths, spells, names and techniques) are all called *kukup* (described to me as 'example'). Whenever one old man touched on important matters, besides his characteristic habit of blowing gently on the thumb and first two fingers of his right hand, he would also feign an incision into his left thigh with his right hand – a common gesture accompanying an exclamation about something that was hidden but has now been shown – and with this he would advise me that this advice would make me strong, and to keep my knowledge hidden in my thigh. After some Kwermin visitors passed through the village, several men used this gesture to convey how the visitors had been impressed by how much activity and how many people there were in Bolivip 'they cut the thigh open and saw us all inside – many and strong!' As a bodily substance kept especially in the thigh muscles, *lamlam* refers to the shiny-membrane covering muscle, to advice and knowledge, and to the shining effects of these when manifested in the world – strong and healthy plants, people and pigs all have a shine to their skin. The sun glinting on a shiny axe is *lamlam*. A person who is walking tall, skin shining, returning from a well tended taro garden is described as *lamlamso* ('with knowledge/water'). 'Knowledge' then, takes the form of bodily substances reserved until a person is ready and would not be harmed, and circulated in exchanges of care. Also conceived of as a kind of bodily liquid (*wok*), this *lamlam* circulates in these exchanges of skin – some gardeners describe how they can also pass *wok* to their taro plants and their children by looking after them and playing with them – as one man put it '*wok* is another word for love'.

When used as 'example', *kukup* is used to describe part and whole of this knowledge and is also used in terms of referring to a person's own 'example' – their acting out a particular combination of examples, akin to personality as Robbins (2004) describes it. This acting out is watched and listened for with attentive care by those who may yet reveal some further *kukup* (in the terms of the cult 'elder brothers' reveal to 'younger brothers'), and who are withholding knowledge until they perceive a readiness and need amongst a more junior grade. Revealed by the way these men talk and act, in comportment, success in gardening, hunting and in problems of marriage, fatherhood and life-cycle exchanges, these capacities and dispositions are perceived in the appearance of skin (*kal*). Ritual activities respond to perceptions of the appearances of people in general (whether their skin is 'light', or 'dull', whether the village is 'hot' or 'cold'), the appearance of the gardens and satisfying qualities of the harvest, and the appearance of a particular initiation grade (perhaps noting their activities, talk and readiness). Equally, hidden revelations of *awem* observe readiness and appearances: men say that if shown too early 'you will begin to look like your older brother'. Similar explanations are given for why one man will decide of his own accord to share a meal of some previously forbidden bird or marsupial with his younger brother: for example, *dubol*, the Dorian's Tree-Kangaroo, has a grey tinge to its fur and appears in myth as an old man – eating *dubol* is held to make one's skin like his and so is avoided by men until middle-age. Young initiates are allowed only birds and marsupials that feed in the canopy (*eil awon, eil nuk*), and are forbidden those that are restricted to the ground and eat worms (*bakan awon, bakan nuk*), lest their own movements become heavy and their skin spoiled by the effects of worms. Marsupials found in a mother's pouch have such soft skin and bones that they are avoided by those looking after their own appearance.

People describe the movement through life and the process of ageing in terms of a journey descending down from the top of a tree – where infants move about from one thing to another as their attention diverts and they follow different paths – to the base where the very elderly remain in the house, with their finger and toenails curling over with little use, unable even to relieve themselves outside. Those in the middle of life are especially active in making gardens – to feed children and those who look to them for support. When looking for a garden site, people inspect not only the types of plants growing there and the soils in which they grow but also examine their relations to the place and persons able to effect growth there, that is, to those who first cleared the forest, or who most recently made a garden there. Gardeners reflect upon whether it was a place that their relatives or friends (*takon*) ever used for hunting or gathering, whether they knew the stones, rivers, trees or marsupial holes there, and remember their own contact with a place. Having established that connection, people then reflect on their relationship with those relatives; whether they were treated kindly by them or looked after them in return, and whether they helped them with their projects or listened to their talk. Having given respect and care in the past, gardeners might feel confident that they can expect to be helped; that their requests for the taro to be kept free of pests, insects and blights, and for the taros to be given 'water' (*wok*), should be acted upon. When harvesting taro, then, people inspect them as if they were artefacts of the social relationships in which they have grown. These plants, then, are imagined as extensions of a gardener and an extension of their efficacy – both in terms of their skill and knowledge, and as evidence of their capacity for respect and care. In some sense, every kindness might one day grow a taro.

Men in Bolivip say that there is only one path to learning important knowledge: one must 'look after the skin' (*kal kiin moyamin*) of someone with renown, 'clean the sleep from their eyes', 'hear their talk', feel and show sorrow for them, fetch water and firewood, give them parts of marsupials and types of taro. Only then will they take equal care in making their advice and *awem* both 'clear' and 'straight' (*turon*). These relations of exchanging skin – knowledge in respect of prior care – are also pursued with a father and mother's brother. Having looked after his skin, and having received his *awem,* one old man became fond of telling me that his skin had gone onto mine, that my skin had gone onto his and that we were now 'one skin'. Through these exchanges and considerations, we had become encompassed by the same skin – what was ours to circulate, circulated within. This path is one that initiated men follow to learn the *awem* of their clan or ground, or *awem* not shown in the initiations or about initiations they have not yet been shown. Women can also follow this path, either to learn *awem* from their father or husband, or in order to learn about important women's knowledge from their female relatives: about techniques, for example, to shape the features of a newborn to make evident the care into which it has been born. Indeed, men give the example of Afek in support of their advice to younger men that they should begin to reveal some pieces of *awem* to their wives – 'they were the first to know'. Some women were also well aware of the men's belief that only vaginal mucus (*abuk gom*) and semen (*iman wok*) are necessary for conception, but they asserted that, in addition, womb or menstrual blood (*abuk kas*) is required. Where the male party regards two things as sufficient, the female party knows that the mixture contains contributions from another source. Both parties agree, however, that having sexual intercourse only once is not enough, and emphasize that repetitions are required.

Men say that a father is the main source of *awem*. He is, of course, keen to make a son capable and strong. Indeed, during initiation beatings, men say that those on the father's side will be the harshest as they try to impart more strength and bravery, and that it is those on the woman's side who will look after an initiate (calling 'your mother is here', 'your sister is here'). Old men say it is sufficient to look after one's father and to ask again and again until one has the stories straight. They disparage the accounts of juniors who tell of having to add things together for themselves: older men simply reiterate that this demonstrates that these people are being tricked by people who do not have things straight themselves. For old men, knowledge is the result of the joint work of a father and son. But the sons tell a different story.

Young men describe the crucial role of having *awem* from another source, notionally from a mother's brother – that this will 'strengthen' that received from a father (the idiom used is *nanor nanor,* which is used to describe the solidifying of drying sago, and the congealing role of conception substances when mixed together). They say that it is possible to hear a story again and decide which one is most complete and straighter, and learn about the bits left hidden or left out. Contributions from a mother's brother, then, are added to the joint work of a father and son. The asymmetry of male and female conception beliefs is also rehearsed by these different experiences and practices in making knowledge. In each instance, an exclusive addition from the mother's side is added to joint work: it is as if 'conception' remains incomplete, allowing for further additions, as though certain capacities of father's side and mother's side are released to grow a person throughout life, and withheld until the person's appearance suggests readiness.

Aside from these exchanges of skin there are other actions that show in a man's skin and his knowledge. When first taken from their mother's house, young boys who are to be shown the *daksal* rite are moved into the *kim am* ('the middle house'), made with walls of twigs and leaves, raised off the ground by branches, and with a roof made of leaves. They are only allowed to eat birds and marsupials that similarly dwell in the forest canopy. Each morning they are taken into the forest and shown how to clean their skin by rubbing with certain leaves and how to make themselves shine by applying pig fat. Several of the initiation rituals have an ambition to release 'water' into initiates, women and children, gardens, pigs and fruit trees. Older men are fond of advising about the benefits of the forest in 'giving new blood' simply through movement and contact with the undergrowth – they say that whenever a man feels lazy or when his skin has become greyed with fire ashes, a visit to the forest will restore energy. Although men make considerable efforts to hide their success in hunting marsupials, they say that there is no hiding the evidence of a recent meal – the blood, especially, helps the skin to glow and shine. Older women say that if their husbands became big-headed with the lightness and strength that their own efforts (in caring for them, gardens and pigs) had afforded them, they used to easily revoke the effect by making an early return to the household from their separate menstrual house (*wok am*) confinement before their flows were through. They say this left their husband's skin dull and heavy, and their ability to wander considerably reduced.

These bodily resources – growing knowledge and repertoire of *kukup*, and the ability to give effect – are all contingent on maintaining good relations with others. Junior cultists have also to learn just how to reveal their knowledge: a misjudged contribution to a discussion in the men's house, where an ethos of heightened respect maintains, will draw remarks intended to shame '*kablam kukup e, e?*' ('is that your own example, your own custom'), with the clear implication that it bears no resemblance to ancestral examples. As one might now imagine, in Bolivip this is only partly an expression about being wrong – often the issue at stake is that a junior, 'younger brother', has acted like an 'elder brother' to his seniors. As such, he has not only to manage a demonstration of knowledge without revealing his ignorance (Borofsky, 1987), but also having to enact his respect for the gender difference between *wanang am alin* and *kinim am alin*. As in other moments, the definition afforded to an interaction and the control of reception lies with more senior men. These divisions also carry the divisions of knowledge, gender and reproduction described above in which the skin of man is treated as an organ of growth, as if it were an analogue of the 'skin' within a woman's womb.

That which circulates between two men encompassed together as 'one skin' in Bolivip must not be allowed to appear 'clear' – improper revelations of important knowledge will attract sorcery and draw suspicions of hidden valuables (*imal kisol*) having been exchanged. Whilst the junior man has to be seen to be respecting seniors (so as to allay suspicions that there has been a hidden exchange of valuables), there is tacit acknowledgement that the senior man will be making returns of knowledge. But the senior man must take care not to reveal 'outside' this containing relation – premature exposure registers on the skin of old-looking juniors and, potentially at least, in harm to the skin of a senior caused by sorcery.

Our junior cultist, then, is able to grow his knowledge and his capacities by locating himself in the scheme through adding himself to various halves – the presentation of juxtaposed imagery, which he may well suspect is only half the story; bringing together the pieces of 'example' that he has been shown and told and has overheard; maintaining good relations in exchanges of skin and in marriage; combining the advice of his mother's brother to strengthen and congeal that of his father; and respecting the internal gender divisions within the men's house, which he discovers are also internal to himself. He is grown by his interactions with others and depends upon paternal and maternal kin to impart further bodily resources. As with harvested taro plants, people inspect his skin for evidence of those responsible for growth and in whom he remains invested.

Echolocation

Melanesian ethnographies have often emphasized the interplay between sound and vision, revelation and concealment and the gender differences defined through secrecy. That women may hear but not see hidden male ritual activities was an early commonplace, exemplified by the 'sexual antagonism' debates (see Poole and Herdt, 1981). Analogies were readily perceived in terms of sexual properties and the possession of secret knowledge, such that quantities of female body substances could be removed from initiates to be replaced by male knowledge revealed during the initiation ritual. Similar concerns with quantity and possession have characterized analyses of knowledge and secret knowledge in particular (see Fardon, 1985). But even within the terms of Barth's own analysis, the Baktaman seem destined to failure in their efforts to transmit a view of culture: 'In the words of one senior: "You know how it is during your initiation: your *finik* (spirit, consciousness) does not hear, you are afraid, you do not understand. Who can remember the acts and the words?"' (Barth, 1975: 101, and see Barth, 1987: 26). As I argue elsewhere, Barth's interpretation of secrecy and his analysis of cultural variation fall down as soon as revelations outside the rituals and to uninitiated men and women are taken into account (Crook, 1999, in press).

In her discussion of the Melanesian aesthetic premise that forms appear out of other forms, Strathern suggests that 'perception is a procreative act' – her example here, is of a 'child being brought forth from the body of a parent (Strathern, 1992: 245) – the perception that something appears as having emerged or been revealed as the inside of something else. For example, a gift makes evident the relations and intentions 'inside' the donor. In Bolivip, we have seen that junior cultists develop relations with a knowledgeable senior man in order to learn the important knowledge that is hidden within him, and develop a perceptual capacity to see 'inside' this knowledge once it has been revealed to him. Their accounts of going inside the men's house for the first time tell of discovering new divisions – the house is divided into two sides, and the body of men is divided between 'male' and 'female' with respect to those who have and have not been shown a particular initiation. Revelations often take the form of juxtaposed images and junior men tell of their frustration and difficulties in deciding which of the possible analogies 'inside' these revelations is the right thought path to follow.

In order to elicit these revelations, a junior man must reveal his own 'insides' to the senior man, by demonstrating generous attention, care and love. This mutual revelation of 'insides' is marked by talk of them developing a shared skin encompassing around or 'outside' of them both as 'one skin'. Once junior men have been shown inside the cult house then, they begin to demonstrate their own inside capacity to more senior men. What returns to them depends on what they give out – echoes depend on location – but also the understandings they form in this important knowledge are a reflection of their existing composition from the 'insides' of other men. In this sense, they share the perception of those whose knowledge has combined with others. That is, they come to share not only skin but also the 'insides' of other men. Their perception, then, is not only their own – the circumstances of the observer are neither objective nor individually subjective. Their world outside mirrors their world inside and comes to reflect the particular interests, concerns and understandings of other people – perception as a procreative act in a supplementary sense. The double ontology evident in the figuring of different positions on a tree show that these 'knowledge making practices' involve different modes and are given a gender cast.

Junior cultists are hardly strangers to arguments among myths: they have to grow a 'skin' around themselves and other men, in which they can grow their knowledge – that is, combine 'examples' and techniques into a version that they can enact. The work of mixing or combining examples with those of paternal and maternal kin such that their 'dizziness' is resolved and straightened is likened here to a woman's work of combining and congealing substances in conception. This ability in knowledge is held and revealed in their own skin, which also acts as a prompt for further resources from those who withhold them until readiness or need has been demonstrated. It is in this sense that the subject-position is defined by what they do with knowledge, rather than by its possession. Faced with juxtaposed examples, advice from different sources, and halves that resolve into one, this can be a dizzying experience. The echolocation of inner capacities requires considerable work.

Wagner's suggestion is that whereas a bat relies on the difference between itself and the objects around as it flies about, human beings make use of the differences between intent in language as a way of locating 'meaning', and locating themselves as a subject: '[T]he bat does not *send* "information" it *receives* it, and the message makes sense only in terms of the relative positioning of subject and object, that is, in those of the bat's motion' (Wagner, 2001: 138). Rather than focusing on visual revelations and emissions in language echoing uniformly – as if independent of the dispositional state or even the position of the originating source – my attention has been caught by the kinds of surface textures and appearances that shape their own echoes in language. Wagner (2001: 138) suggests that 'A bat that flew at Mach 1 would be in deep trouble', and I have suggested that acting beyond the limits of their own internal composition can spell deep trouble for the growth of male initiates and their taro plants. Just as contact with a cave wall will harm a bat, contact with important knowledge before they are ready will harm initiate and plant alike.

Doubly Special Relativity

The developments and transformations stimulated by relativity thinking in physics and in anthropology were, of course, far from straightforward and, as Stocking (1996: 712–13) puts it 'the diverse discourses that may be historically subsumed by [the term 'anthropology'] have only in certain moments and places been fused into anything approximating a unified science of humankind.' In his discussion of Einstein's 'wake' or bearing on modernist literary culture, Whitworth (2001) emphasizes that ideas drawn together by relativity had already provided new descriptive metaphors and were also being assimilated into established concepts. In anthropology's 'formative crises and transformations' then, there was nothing that resembled a ready-made template for the transposition of metaphors. But arguments are made with and against something, and it is telling that a 'science of humankind' should perceive the argument in terms of interacting with and fashioning itself against the natural sciences such that its own questions and discourse carried the concerns of another project (Viveiros de Castro, 2003). Paradoxically, arguments about relativity in anthropology – against social evolution and biological determinism as 'physical laws' or against objective scientism – also reinforce the notion of there being something to be relative to. In the arguments among myths, between metaphors of 'relativity', anthropology has acquired definition as a subject through the echolocation against other subjects including physics. In this sense, anthropology is only half its own story.

Unlike the variation that comes of different beings occupying the subject-position of observer and seeing the 'same' thing in Viveiros de Castro's example of Amazonian perspectivism, the materials discussed here suggest that age, sex, gender, aptitude and respectfulness each inform the variable subject-positions of the observer. That is, the observer is not the same – what comes back to people in Bolivip depends on where they are – echoes depend on location. As Feld (1994: 9) puts it when describing 'acoustemology' (acoustic epistemology) as 'a special kind of knowing', 'echo is about presence, about reverberant pasts in the present, presents in the past.' But among the pitfalls that renewed interest in the anthropology of *knowledge* must avoid is an impulse to privilege 'knowledge' – as some pan-human instrument or culturally relative information set – over those doing the knowing. Idioms of 'knowledge' as some kind of universal 'method' through which humans live and orientate themselves in the world – 'what a person employs to interpret and act on the world' (Barth, 2002: 1) – carry a curious similarity to the Academy's own interests in a disciplined method for knowing, producing and coming into possession of facts about the world. Routing connections and relations through people is second nature to anthropology: persons know through other persons, and knowledge is always known in relation to something else. But if our own metaphors of knowledge are coupled with ideas of kinship (Strathern, 2005), then how we think about knowledge, persons and relations will require similar analytical sophistication.

Perhaps if anthropology were to develop a doubly special relativity of its own, it might absorb the new epistemological metaphors afforded by a variable speed of light and do so in terms of its existing concerns with the multiplicities of meanings deriving from the circumstances of the observer. But it might also question the division and separation between insides and outsides (see Gell, 1998: 140–1) which has lent some of our examples the character of only taking into account a physical location, a subject position or a sociological position in the 'outside' world and pretending that it is the same for everyone, whoever they are. As we have seen, in Bolivip where ontology and gerontology ideally keep pace with each other, the 'inside' disposition and composition of the person is not so easily demarcated from their 'outside' position in the world. Wagner's image of echolocation also collapses the division, making the subject's variable definition a feature of relations in language. In short, it matters very much who it is occupying the position. In a world where knowledge flows as a substance between people, from ritual and into plants, the skin becomes more than the surface of a division or site of separation – it receives, absorbs and grows a person's echolocation in others. The image of a tree contains both the branching possibilities *and* the possibility of 'one story' – as if relativity and rationality in this scheme were different positions in the same thing rather than mutually excluding perspectives. Anthropology is doubly sensitive to inside and outside positions – its own version of doubly special relativity, then, might recognize that it is not enough to simply occupy a subject-position without also considering the internal state of the occupier and the relations in knowledge, language and kinship through which this echoes and is located.

Acknowledgements

The research for this chapter has received support from many sources: the ESRC and RAI/Sutasoma Award funded initial fieldwork in Bolivip, whereas the British Academy provided postdoctoral opportunities, as did my involvement with those engaged in the Cambridge-Brunel ESRC project 'Property, Transactions and Creations'. I remain grateful to these bodies for their interest and assistance, as I do to the Papua New Guinea National Research Institute for allowing me to affiliate myself with them. I hope I continue to make my acknowledgements of the part of myself that was given me in Bolivìp to the satisfaction of those who helped me, particularly Dominicus Sulumeng who was so generous in giving, and looking after my skin. Hoenik Kwon put the ideas for the perspectivism panel in my head and grew them through discussion, not only in Manchester where Eduardo Viveiros de Castro and Marilyn Strathern also reminded me of how other ideas here had first sounded. Versions of this chapter were given in St Andrews and Brunel: my thanks to Mark Harris and Eric Hirsch for their stimulus. Adam Reed's advice and Christina Toren's encouragement were particularly timely. My thanks to the combined editorial talents and patience of this volume's editors in their readings and comments as the chapter resolved itself under their guidance.

Bibliography

Barth, F. (1975), *Ritual and Knowledge among the Baktaman of New Guinea*, Oslo: Universitetsforlaget.

—— (1987), *Cosmologies in the Making: A Generative Approach to Cultural Variation in Inner New Guinea*, Cambridge: Cambridge University Press.

—— (2002), 'An Anthropology of Knowledge', *Current Anthropology*, 43(1): 1–18.

Barnes, B. and Bloor, D. (1982), 'Relativism, Rationalism and the Sociology of Knowledge', in M. Hollis and S. Lukes (eds), *Rationality and Relativism*, Oxford: Basil Blackwell.

Bateson, G. (1958 [1936]), *Naven*, 2nd edn, Stanford, CA: Stanford University Press.

Bercovitch, E. (1989), *Disclosure and Concealment: A Study of Secrecy among the Nalumin People of Papua New Guinea*, PhD Thesis, Stanford University.

Borofsky, R. (1987), *Making History: Pukapukan and Anthropological Constructions of Knowledge*, Cambridge: Cambridge University Press.

Crook, T. (1999), 'Growing Knowledge in Bolivip', *Oceania*, 65(4): 225–42.

—— (in press), *Anthropological Knowledge, Secrecy and Bolivip, Papua New Guinea: Exchanging Skin*, London: The British Academy.

Eggertson, S. (2003), *Transforming Skins: Kwermin Sequences of Growth*, Reykjavik: Iceland University Press.

Fardon, R. (1985), 'Introduction: A Sense of Relevance', in R. Fardon (ed.), *Power and Knowledge: Anthropological and Sociological Approaches*, Edinburgh: Scottish Academic Press.
Feld, S. (1994), 'From Ethnomusicology to Echo-Muse-Ecology', in *The Sound World of Bosavi*, Sleeve booklet accompanying Bosavi CD set.
Gardner, D. (1983), 'Performativity in Ritual: The Mianmin Case', *Man*, 18: 346–60.
Gell, A. (1998), *Art and Agency: An Anthropological Theory*, Oxford: Clarendon Press.
Gillison, G. (1993), *Between Culture and Fantasy: A New Guinea Highlands Mythology*, Chicago: Chicago University Press.
Herdt, G. (1990), 'Secret Societies and Secret Collectives', *Oceania*, 60: 360–81.
Jorgensen, D. (1990), 'Secrecy's Turns', *Canberra Anthropology*, 13(1): 40–7.
Lipset, D. (1980), *Gregory Bateson: The Legacy of a Scientist*, Englewood Cliffs, NJ: Prentice Hall.
Ortega y Gasset, J. (1931), *The Revolt of the Masses*, New York: W. W. Norton & Company, Inc.
Poole, F. J. P. (1982), 'The Ritual Forging of Identity: Aspects of Person and Self in Bimin-Kuskusmin Male Initiation', in G. Herdt (ed.), *Rituals of Manhood: Male Initiation in Papua New Guinea*, Berkeley: University of California Press.
Poole, F. J., and Herdt, G. (eds) (1981), 'Sexual Antagonism, Gender and Social Change in Papua New Guinea.' Special issue of *Social Analysis*, 12.
Robbins, J. (2004), *Becoming Sinners: Christianity and Moral Torment in a Papua New Guinea Society*, Berkeley: University of California Press.
Russell, B. (1925), *The ABC of Relativity*, London: Kegan Paul, Trench Trubner & Co.
Stocking, G. (1996), 'Paradigmatic Traditions in the History of Anthropology', in R. Olby, G. N. Cantor, J. R. R. Christie (eds.), *Companion to the History of Modern Science*, London: Routledge.
Strathern, M. (1992), 'The Decomposition of an Event', *Cultural Anthropology*, 7(2): 244–54.
—— (2005), *Kinship, Law and the Unexpected: Relatives are Always a Surprise*, Cambridge: Cambridge University Press.
Viveiros de Castro, E. (1996), 'Cosmological Deixis and Amerindian Perspectivism', *The Journal of the Royal Anthropological Institute*, 4(3): 469–88.
—— (2003), *AND*, Manchester Research Papers, Social Anthropology, University of Manchester.
Wagner, R. (1975), *The Invention of Culture*, Englewood Cliffs, NJ: Prentice Hall.
—— (2001), *An Anthropology of the Subject*, Berkeley, CA: University of California Press.
Whitworth, M. (2001), *Einstein's Wake: Relativity, Metaphor, and Modernist Literature*, Oxford: Oxford University Press.

–10–

Being Human in a Dualistically-Conceived Embodied World

Descartes' Dualism and Sakais' Universalist Concepts of (Altered) Consciousness, Inner-Knowledge and Self

Nathan Porath

Introduction

The knowledges that the discipline of anthropology describes and represents in monographs and particularizes as other peoples' cultures are knowledges that have their own claims to universalist knowledge status. When I ask a Sakai shaman about his ideas about personhood and self, his answers are about the frames of my own internal constituents that make me human as much as that of any Sakai individual. What a shaman or any other Sakai provides as answers to my anthropological queries is presented as knowledge about what makes us human (manusio). Within Sakai society and under the local healer's authority, our cultural differences are mediated by the local ethnopsychology's universalist frame of what being a conscious human being is. Moreover, the local ethnopsychological theory even accounts for cultural difference within its own explanatory discourse. Epistemologically, ethnopsychologies are universally framed, thus making the prefix 'ethno' rather awkward. Through this universal frame of reference (that we are all consciously human), people from another culture can understand and phenomenally even experience the suggestive ethnopsychological metaphors in their very body-consciousness. Such an experience becomes the methodological basis of anthropological fieldwork for obtaining knowledge of 'others' in this area of study (see Fernandez, 1986). The psychologically universal and particular seem to collapse through this intense moment of human coevalness (to use a term from Fabian, 2002; see also Cohen and Rapport, 1995).

Sakais are a group of cassava shifting-cultivators who until recently were a non-monotheist forest-dwelling people speaking a dialect of the Malay language. They live in the upstream-Mandau area of the Indonesian province of Riau (Sumatra). Their area was peripheral to the semi-independent Muslim-Malay kingdom of Siak Sri Indrapura, which after a period of Dutch indirect-colonial rule, was incorporated into the Republic of Indonesia in 1945. In the past, the forest people of the area never called themselves Sakai, a catch-all term in the Malay speaking world for forest-dwelling peoples, which had derogatory connotation. Whereas in Malaysia the term was officially abolished on indigenous request, in Indonesia, this exonym has become the official ethnic name for the indigenous people of the upstream-Mandau area who today are also Indonesian citizens. Since the 1970s, Sakais were exposed to Indonesian state development projects and monotheization. Today (ethnographic present late 1990s), although a small minority practise Islam through Sufism, most Sakais nominally profess the Islamic faith.

In the following, I would like to compare Sakais' views about the human self and embodied knowledge in relation to Descartes' dualist philosophy, the very epistemological foundation of modern Western universal theories of knowledge. Many anthropologists have criticized Cartesian dualist approaches suggesting that we should do away with the distinction of mind and body and instead work with a unified theory of embodiment, which collapses the two. It is argued that Cartesian dualism has been a theoretical blinker within anthropological understanding of other cultures (Scheper-Hughes and Lock, 1997; Csordas, 1990). Authors have also suggested that dualism and embodiment approaches are methodologically incompatible (see Lambek, 1998).

I will describe Sakai concepts of human consciousness in relation to this critical interface between Cartesian dualism and the anthropological concept of embodiment. From the ethnography most peoples do experience and perceive the world through a dualistic framework. I shall therefore explore Descartes' own philosophy of mind and body in relation to Sakais' dualism. We should not simply force our ethnography into a monist frame when the facts tell us otherwise. What should be stressed, though, is that although Sakais are dualist, their dualism should not be confused with mind/body dualism. But neither should it be reduced to the 'mind in the body' construct that may be another semantic version of this dyad. Sakai dualism is based on the materiality of a physical and non-physical distinction, which as Tylor pointed out long ago (although not in these exact words or implications) has developed through the human phenomenal experience of altered states of consciousness. Rather than starting with the premise of conscious awareness, Sakais' approach to consciousness starts with the dispositional experience of altered states of consciousness.

The question that this chapter tries to resolve ethnographically is the relationship between Sakai dualist experience and embodiment experiences – how do they connect. To answer this I will also briefly explore Sakai hypnotic realities and their experiential theory of the embodiment of knowledge to induce hypnotic realties (what in general anthropological literature is referred to as 'magic'). Through the ethnography I show that methodologically dualism and embodiment (monism) are commensurable. The pre-requisite for embodiment is a conceived dualist experience of consciousness.

It might be argued that making a critical comparison between the ideas of an indigenous culture and one seventeenth-century European thinker is inappropriate as we are not comparing like with like (see Lambek, 1998). Although Descartes was a man of his time, and was still influenced by the scholastic philosophy that came before him, his writings harbour the cultural grain of the directions that European scientific knowledge was moving towards and thus he laid the foundations of its structural shape. This is not to say that all scientific thought has been dualist since Descartes. However, monist traditions in the different disciplines have been critically trapped within the Cartesian dualist's net (see Midgely, 2001). Each discipline in the social sciences has formed its own sub-dualist/monist traditional dichotomies (Lambek, 1998). These scientific traditions have not only shaped the modern societies they emerged from but have determined the way the West came to view people of the rest of the world as these modern societies politically and culturally interacted with them. I feel that indigenous peoples' knowledge should be brought directly onto this conversational band-wagon, as Descartes' ultimate onslaught was on these very kinds of knowledges, which he implied as being nothing more than 'confused thought'.

An Anthropological Moment with Descartes

From the opening of Descartes' *Discourse on Method* (hereafter DM, 1996), it becomes clear that he was aware of what today we would call cultural and traditional diversity. The question that prompted his method of universal doubt rests on what anthropologists could recognize as a fundamental epistemological-anthropological question: If humans are the product of their cultural and educational upbringing, then how can we ever attain universal certainties of knowledge about the world that are valid for all humans? On reading the opening of DM, one gains the impression that Descartes sees his attempt as a quest for salvation out of the chaos of diversity. Descartes' writings use the religious narrative of the times (Pavel, 1996: 358). Although his writings resonate with religious overtones, the salvation he is looking for is the salvation of knowledge of the world through Reason (DM: 35). Descartes aim was to find a means of overcoming the 'particular' and find some basic knowledge that could provide a method for universal human truths (Toulmin, 1996: 125). His universalistic-humanist approach justified itself on the principle that all humans, no matter which society they are from or whatever their upbringing may be, are endowed with Reason and the potentiality for it. The basic propositions he searched for to base his theory of knowledge were those logically structured like geometrical truths such as 'a square has four sides'. Such statements could never be otherwise even in a dream. For Descartes, the search for these kinds of logical propositions was the real activity of a rational thinking mind.

Descartes' approach to the problem of knowledge in his *Meditations of First Philosophy* (hereafter MFP) was through 'methodological doubt' (MFP, 1996: 58). He raised a doubt on everything that he knew, including God, suggesting that maybe a demonic evil genius was deceiving him into believing that all that he thought he knew to be true was in fact mere illusions similar to the images experienced in dreams (MFP: 61). Descartes' methodological doubt led him to a logical conclusion that no matter how much a demon can deceive him, he can never be deceived into thinking that he does not exist, for one has to exist to be able to doubt, to be deceived, to think – hence 'I think therefore I am' (MFP: 62). This indubitable 'thinking thing' is Descartes' mind or soul.

At this point in his philosophy, Descartes' 'soul' existence is like a thinking ether that has still to prove to itself logically that it is not the only thing in existence, which is deceived on everything else that it experiences. Descartes concern was not with the phenomena, but to prove that things really and truly external to him (the object) cause the phenomenon of perception. What made the *cogito* so certain was that it was clear and distinct. This he then took as a general methodological rule (Russell, 1946: 548).

Once Descartes' certainty of a thinking self had been established, he tried to restore the logical certainty of the existence of the rest of the world. Descartes next step then was to prove the existence of God, which he does through scholastic arguments. God provided Descartes with the path out of his solipsism, and allowed him to tread, albeit carefully, on the path of knowledge and understanding of the world external to the thinking and experiencing mind (MFP: 95). Logically, Descartes never managed to pull himself out of his 'soul' existence, as his arguments for the existence of God begged the question he was asking. Descartes may have had ideas about an all-perfect being but these were part of the 'illusions' that he doubted unless they could be logically proven.

Nevertheless, feeling secure in his logical propositions, Descartes proceeded to explore the existence of matter or body. Descartes defined matter as an extended object that occupies space, and is fundamentally distinct in substance from mind and from God. Once having created a dualism between spirit/mind and body/matter, it was imperative to explore the relationship that these two distinct substances have to each other. The human body, which is composed of matter has certain needs, Descartes called 'animal spirits'. These needs are its reflexes, what it does unthinkingly, and these 'animal spirits' humans share with animal bodies. The most critical question of Descartes' philosophy was how does the mind connect to this body (Strathern, 1996: 6). Descartes postulated that the mind is united with the body through the pineal gland (MFP: 120) and bodily sensations are the product of the intermingling of the two substances.

It is through the connection of mind to body that the mind's rational experiences are distorted, providing the type of chaotic diversity of knowledge that Descartes' methodological doubt tried to overcome (MFP: 107). Descartes concluded that it is mind alone and not the combination of mind and body that is necessary for understanding truths and obtaining rational knowledge about the world (MFP: 103). In order for the mind to achieve its fullest rational potential, he believed that the ethical life should exhort extreme moderation with regards to the body. The mind had to remain superior to the body-object

For Descartes, people can clearly and distinctly conceive physical objects and ideas. He acknowledges that humans also have the experiences of dreams and can hallucinate. However, he did not develop a theory of dreams (and other altered states of consciousness) in which people also experience people and objects as real, other than suggesting that these impressions are the product of the mind's embodied relation with the baser body and, from the point of view of universal knowledge, are false images and illusions. They are phenomenal experiences of the mind's interaction with the body, with no real or important existential cause. In his philosophy and subsequent Western philosophy, the experiences of altered states of consciousness had a negative status in human experiences (Noll, 1983). No knowledge of the objective world could be learnt from these aberrant experiences of the mind. These experiences were subsumed to reason. Knowledge from the altered states of conscious experience (including madness) had nothing of value to say (Foucault, 1965).

Oddly for anthropological embodiment theory, which has developed as a critique of Cartesian dualism, Descartes' mind/body interaction is itself a theory of an embodied mind. Descartes stressed that the mind did not exist in relation to the body like a pilot governing a vessel, but is closely related to the body 'that I [here meaning the mind] seem to compose with it [here meaning the body] one whole' (MFP: 120). Strathern (1996: 42) points out that Descartes' mind/body interaction suggests a third realm that comes closer to Merleau-Ponty's philosophy of the body. Merleau-Ponty (1962: 369) himself also seems to suggest that his philosophy, which is a critique of scientific dualism, is fundamentally engaged in a critical dialogue with Descartes. The problem seems to be that Descartes disparaged the experiences that are produced through what he saw as the intermingling of mind and body calling them 'confused thoughts' (MFP: 102). These 'confused thoughts' are the very experiences that the study of cultural praxis, phenomenological anthropology, and ethnopsychology is concerned with. Descartes' 'confused thoughts' have become the subject matter of the social sciences and particularly anthropology.

Sakais' Fundamental Dualist Ontology

For the Sakais of Riau the fundamental ontological dichotomy is of a physical dimension and a spirit dimension, *a fundamental dualist dichotomy* that cannot be denied to their worldview (see also Benjamin, 1979). The physical dimension is called the human dimension (*alap manusio*). Although the determining factor is its physicality, humans always perceive it from their perspective; hence it is the human dimension. Categorically incorporated within the human dimension is the forest, animals and plants – that is, everything that humans can perceive with their five senses.

In contrast to the human dimension is the other dimension (*alap lain*). The spirit dimension cannot be perceived with the five senses, and the characteristic of its materiality is non-physicality. This dimension exists asymmetrically parallel (*te'balik*) to the physical one and both these dimensions equally constitute this world (*dunio iko*).

At night, when the *alap manusio* (human-dimension) is asleep, the spirits in the *alap lain* are awake and carry out various activities that humans do during the day. Everything that exists in the human dimension also exists in the other dimension but in a more perfect ideal form. The people here are more beautiful than their physical counterparts. They own boats, ships, planes, cars, and a myriad of consumer goods that exist in the human dimension, but occur in more perfect form in the other dimension. These goods do not exist as materially physical objects. Their perfection lies in the nature of their existence as refined images (*mayo*). The boat is a singular image (*mayo*) and, as an image, it is both imagined and conceptualized. Moreover, in the *alap lain* there are as many images (*mayo*) of boats and other goods as there are spirits who own them. Compared to their physical counterparts, these images are superior in the quality of their material refinement (*bahan alus*). The spirit dimension is therefore a materially refined imag(e)inal dimension. People can see (*nengot*) the imageries (*mayo*) of this dimension with their inner-eye of consciousness (*mato batin*) when they lose consciousness of their physical environment (*tak soda' lai* – not aware anymore). This unmarked expression *tak soda'laí* refers not just to a particular experience of what we call an altered state of consciousness. It is a local expression reflecting recognition of generic experiences of consciousness in which the individuals lose their awareness of the physical surroundings.

The Person, Self and Interaction with the People of the Spirit Dimension

Among Sakais, the word for person (and people) is *o'ak* (or *o'ang*). It is also the word that people use to define geographical and ethnic group membership. In

Sakai thought, so it seems, an individual does not necessarily belong to only one group, and can be categorized in various ways according to context. Identities are not definite but conceptually and experientially fluid. A person born from parents from the upstream Mandau area, but brought up and adopted by parents elsewhere, is considered to be a person who, although born a 'local offspring' (*anak buah*), has nevertheless become a person of his/her associative group (see also Astuti, 2001). Likewise, a person associating with spirits is close to being and becoming a spirit-person too.

An *o'ak* is a fully conscious and wilful social being. Other than humans (*manusio*), spirits are also a category of *o'ak* (people) regardless of what form it might take (animal or human). To hear or see a spirit is to socially interact with one, and therefore they are people (see also Skeat, 1965; Howell, 1984). As a category of people, spirits are usually positioned as 'outsider peoples' and the normal shamanic relationship is portrayed in patron/client socio-relational terms. A lone individual is vulnerable to interaction with these *o'ak*, which can lead to illness, madness and even death. But how do spirits socially interact with people? For this we have to explore Sakai ideas of self and human consciousness.

The Sakai word for self is *di'i*. Unlike Descartes' concept of the self, which is clearly related to the mind, the *di'i* of a human person (*o'ak manusio*) is a composite of a number of elements: the body (*badat*), breath of life (*na'o*), the inner 'storehouse' of personal experiences, knowledge, dispositions and psychic potential (*batin*) and the liver-heart as the seat of emotions and feelings (*'ati*), reason and intellect (*akal*), the soul in an non-experiential metaphysical sense (*jiwo*) and what is known as *semanget.*

The body is an unrefined base substance (*bahan kasa'*), in contrast to *semanget*, which is a refined invisible substance (*bahan 'alus*). The *jiwo* is an Indic derived-term for soul (Yousuf, 1994: 4). Every individual has his or her own *jiwo*. However, the *jiwo* is not something that people consciously experience unless they verbally equate it with *semanget.* The concept of *jiwo* does not feature in shamanic healing sessions either. It is an idea about a metaphysical soul, which the forebears of the present-day Sakais (the people of the upstream-Mandau area) adopted in the distant past from the outside but did not develop in an experiential way.

As long as the body is infused with the breath of life (*na'o*) it is alive (*masih 'idup*). But the life of a body is determined by the embodiment of *semanget*. Every living creature that can move has *semanget*. A foetus already has *semanget* when it kicks in its mother's womb. Sakais associate *semanget* with the blood running through the veins. Its presence is felt at pulse points, for example the wrists, the chest and the forehead. These points are its 'seats' in the body. A body (*badat*) that does not have *semanget* is a dead body. Once humans or animals lose (all) their *semanget,* they also lose their *na'o* (breath of life) and die. The lifeless body begins to decompose and loses the image (*mayo*) it received from its embodied *semanget.*[1]

During a person's life, *semanget* can frequently wane by partially and momentarily detaching itself from the body. This is referred to as '*semanget* taking flight' (*semanget to'bak*). The disembodied *semanget* is metaphorically characterized as a tiny firefly that can somersault freely in the air and fly with imperceptible speed. The more common metaphor is that of a bird. In shamanic healing sessions, the disembodied *semanget* is personalized as the *semanget* bird (*bu'uk semanget*). As an airborne being, it is timid and can suddenly depart from the body under extreme confrontations. People, and especially the elderly who are easily startled, can suffer from *semanget*-flight. In shock (*semenget tekojit*) the *semanget* takes flight (*tobak*) from the body, leaving the body momentarily unconscious. If *semanget* is prone to fly from fear, it can also be frightened back into the body. A sudden loud shout, or a loud authoritative command to return (*semanget balik*) can restore the 'bird' to the body, and hence cause a return to consciousness.

Semanget exists like a shadow (*bayak bayangan*) that cannot be grasped; it is materially non-physical. It exists like the images humans see in a dream, which are other people's *semanget* entering a person's dream consciousness. The experience of *semanget* gains its reality through the sensation of flight, travel and communication with lifelike, colourful and sometimes frightening shadowy images (*mayo*) in the states of altered consciousness. *Semanget* is the experience of self-agency in those states. Spirits interact with humans through embodied *semanget* and therefore, for Sakai, spirits, like embodied *semanget* are experientially real. This interaction is possible because embodied *semanget* and spirits (and *mayo*/image) are constituted of the same *semanget* substance, which is contrasted to physical matter.

Although Sakais divide the world into two basic realities, when people dream, their *semanget* does not usually enter the spirit dimension. Sakai dualist discourse about the human and spirit dimensions seems to suggests an intermediate third dimension – the *semanget* dimension that connects the two. The two dimensions, the human physical dimension and the spirit dimension, are graded spheres that connect through a grey liminal area of the *semanget* dimension – the bridge between the spirit dimension and the physical human one. This liminal area defines the boundary zone between the two dimensions.

When spirits wilfully interact with people outside of ritualized shamanic context (either in dreams or in the waking state), they break the boundaries between the human and the spirit dimension by confronting their embodied *semanget*. When this happens, people can suffer from physical and mental illness. A spirit that interacts with human-embodied *semanget* usually presents itself to that person to the exclusion of all others. Although humans experience the spirit sighting with their eyes, it is generally understood that they see the spirit with their inner-eye (*mato batin*). Thus like dreams, spirit-sightings (vision) are personal. Spirits also interact with humans through *semanget* while they are sleeping. A spirit takes fancy to a person's *semanget* and tries to entice it to follow to the spirit dimension. Such spirit/*semanget* interactions call for shamanic healing intervention to restore the afflicted *semanget* in the body and keep the boundaries between the two dimensions in place.

Due to the fact that the world is ontologically dual, and one aspect of it, *semanget* is connected to body (*badat*) giving it consciousness, but shares its substance with the matter of spirits who can interact with it, it follows that human consciousness is susceptible to such interactions – human consciousness is frail. Madness (*gilo*) suggests that a spirit interacts with embodied *semanget* causing the person to go mad (*gilo 'antu*). The ritually uncontrolled sighting of a spirit-child, for example, is not a symptom of a person's madness; it is a cause of madness. The spirit-child interacts with a human's embodied *semanget* allowing her to see it. Commonly, the spirit is described as flirting with embodied *semanget* when the victim is awake. The spirit breaks the dimensional boundaries causing the spirit dimension to 'enter' (*masuk*) the individual's conscious awareness. The person's *semanget* also enters the spirit-dimension but in the physical-dimension it begins to behave strangely. Such experiences of altered states of consciousness, however, are not placed within a subjective and objective dualism. Such an experience is both personal (subjective) and is an experience of the 'insides' as much as it is caused by something external to her (object). The phenomenal experience is taken for what the experience itself suggests: a socially interactive experience that has pulled an aspect of the 'conscious self' towards the offending other in a shared space of consciousness.[2] One's conscious state and interactive social experience form a unity.

In the mad state, the madness is likened to a human being locked in a house, and only through shamanic intervention can a 'key' be found to open the door. For Sakais, such spirit/*semanget* interactions are not of a religious nature in any sense, but are seen as dangerous (*bahayo*) and pathological (*sakit gilo* – 'ill mad'). They have to be shamanically averted and/or therapeutically controlled.

The cultural epistemology is intellectually and sensuously structured such that spirits are a fact of life and part of the human behavioural environment (Hallowell, 1955). In relation to humans, spirits can be unpredictable people who cause wanton mischief as they break the boundaries of the two-dimensional divide through the embodied intermediary third dimension of *semanget*, and interact with the human self. If *semanget* is the experience of self in the altered states of consciousness, spirits are the strange experience of otherness in those states. These types of experiences between self and other in the altered states of consciousness justify the ontologically dualist worldview.

A Culture of Embodiment

The culture of the upstream-Mandau area can be described as a culture of embodiment. By this I mean that it is not only an embodied culture but it is a culture based on knowledge about the use of the conscious body in storing and retrieving knowledge, and the use and effect of this knowledge in interaction with others. Moreover, people perceive this knowledge as embodied and today they contrast it with the 'knowledge of reason' (*ilmu akal*), which embodies knowledge in books. As people say, when they obtain knowledge they *tulis di 'ati* (write it in their hearts) rather than *tulis di dalap buku* (write it in a book). Acquiring knowledge like this was the only way a mobile people could maintain it and pass it on (see also Howell, 1984). Embodied knowledge is perceived as emerging from people's 'inners' for temporary materialization and extemporization.

The word for 'inner' is *batin*, and was adopted from Sufism probably during the mid-twentieth century. The original word for the 'inner' is *'ati*, or liver-heart. When talking about *batin*, people always indicate its presence in the abdominal region. One feels (*'aso*) and intend action with one's liver-heart and those feelings are harboured in the *batin*. Feelings (*'aso*) emerge from the liver-heart filling the body up to the eyes and is released through them governing words and actions. The *batin/'ati* is the inner subjectivity or total psychological complexity that makes a human person. It is also the liver-heart where decisive moral intentions of a person's act are generated.

When a person closes her eyes and projects her inner consciousness inwards to an awareness of thoughts and feelings she is introspecting her *batin*. The content of the *batin* is seen with the *mato-batin* (*mato*/eye/*batin*/inner). *Batin*-introspection, is not an altered state of conscious experience. It is an ordinary conscious activity of self-reflection. People simply lie down on the floor of their house and place one arm firmly over the eyes so the eyelids are shut, and then look inwards. From the Sakai perspective, we can say that when Descartes introspected his thoughts, he was introspecting his *batin*.

The *batin/'ati* is the body's inner-receptacle, which receives and stores the person's experiences of the external world. It is the locus of the person's experiences, memories and (cultural) knowledge. The *batin* contains oral and cultural knowledge of the past such as legends and stories and tales. Bards and other storytellers carry memorized words in their *batin* that they can recall and extemporize and also have the inner knowledge to memorize and improvise on these words. The *batin* is also the locus of the shamanic and magical knowledge that gives the individual the potential to effect the world through the unseen *semanget* dimension of it. Shamans/medicine (wo)men have exceptional *batin*, which contains a store of magical spells, songs and quatrains through which they call spirits, affect *semanget* and people's emotions. The average local person's *batin* harbours quatrains, few spells, some songs, knowledge to produce material culture, laymen's knowledge of legends and knowledge of how to survive in their environment (ecological knowledge). It is possible for a shaman not to have a powerful magical *batin*, whereas a non-shaman with a powerful magical *batin* lacks the knowledge and experience to enter the spirit-dimension. Every human being has a *batin*, but some *batin* are more powerful than others in certain respects according to the knowledge they have acquired, and this is sensuously experienced when in their authoritative presence, and when they put their knowledge to use.

Although Sakais do not have a word corresponding to the word 'mind' the word for 'thought' is *piki'an* but in many respects associated with feelings (*'aso)* (for the Balinese, see Wikan, 1989). The thoughts and feelings that humans harbour in their *batin* determine their posture, and what people do with their bodies in interaction with others. Suggestive body communication is crucial and played upon. In interactions with others the released embodied emotions create an atmosphere. Others sense the atmosphere in what we might call the sixth sense. This sense (*'aso*) is the total body sense that is attuned to interactive atmospheres and other people's intentions (both human and spirit). Moreover, people are conceived as matching (*'aso cocok*) or not matching (*'ao tak cocok*) with each other through the sensualities of the embodied contents of their *batin/'ati*. This is also crucial for healing because not all healers' embodied knowledge match with their patient's embodied needs. When the *batin/'ati* of two people does not match (for personal or cultural reasons), people say that the other has a *batin/'ati bu'be'do* (a different 'inner'/liver-heart). Sakai thinkers would therefore agree with Csordas who states that embodiment is the locus of culture (1990) although they would use the word knowledge (*ilmu*) rather than culture.

How does Sakai Dualism and Embodiment Theory Come Together in Sakai Thought and Experience?

Cartesian philosophy would have disparaged the experiences that people have in the altered states of consciousness. Although as phenomena they are experientially real, they have no objective basis in reality. Descartes considered them to be the consequence of the mind mingling with the body. For Descartes, what was of relevance to knowledge was the perception of the objective material world that presented itself to the senses. The questions that Descartes raised about the objective reality of the physical world would be ridiculous to Sakai thinkers. There is no problem about conscious awareness (*sada'*) in the ordinary states of consciousness. Here they would agree with Descartes final conclusion that 'I ought to set aside all doubts of these past two days as hyperbolic and ridiculous' (MFP: 108). For Sakais, the problem of consciousness starts with the experience of the altered states of consciousness, when one loses one's awareness of one's physical surroundings (*tak soda' la'i*).

Like Descartes (and unlike Freud), Sakais would agree that everything that one harbours in one's *batin* is available for conscious inspection by the inner-eye (*mato-batin*). The inner-eye can 'clearly and distinctly' perceive what is presented to it – thoughts (*pikiran*), feelings (*'aso*) and images (*mayo*). Under certain conditions, the *mato-batin* (inner-eye) can also be opened to the spirit dimension. Unlike Descartes but like Freud, the images and experiences that a human individual has in dreams and in other altered states of consciousness are replete with meaning that are important for the individual who experiences them. Moreover, it is an interactive experience that, if managed correctly, can not only provide knowledge but transform the person's 'inner' self (*batin*).

In this fundamentally dualist world, a person's knowledge can have two sources; the human dimension and the spirit dimension. Knowledge drawn from the human dimension is learned by imitation, doing and improvisation. Knowledge from the spirit dimension is usually inner-knowledge (*ilmu batin*) that is obtained and stored in the *batin*. It emanates from the liver-heart and can be used to affect the invisible *semanget* dimension. Spirits (sense of other) convey this knowledge to humans via embodied *semanget* (sense of self) in their dreams. There is a variety of inner knowledge that is thus obtained. For example, if people wake up in the morning after having a lucid dream in which they experience themselves being taught a string of words or a song, then they should make sure that they memorize these words and keep them secretly in the *batin*. Shamans receive songs from their tutelary-spirit in this way. What we call magic (hereafter inner-power-knowledge) through which people influence the *semanget* of others, also finds its source in the spirit-dimension.

When a person wants to learn inner-power-knowledge of this sort, s/he approaches a medicine (wo)man and formally asks to be taught this knowledge (*buaja ilmu*). The medicine (wo)man consults his/her spirit-teacher (*gu'u*) in a dream and once accepted, the initiate usually moves into the medicine (wo)man's house and can start learning the knowledge. The physical aspect of learning inner-power-knowledge is to memorize the spells (*monto*). The initiate has to reiterate (*meng'opal*) the spells at night before going to sleep and when doing various chores (usually for their teacher) that do not require much thinking. The spells are 'inscribed in the heart' (*tulis di 'ati*) as all of Sakai verbal art is. After the initiate has memorized the spells and the teacher is satisfied, the initiate has to give the teacher a payment of a number of symbolic items. The individual can then start using the knowledge.

Much of non-shamanic inner knowledge is manifest in spells (*monto*). These spells are memorized formulas. When a person activates this type of inner-power-knowledge, an unseen force is generated much as a hammer blow can drive a nail through wood. The cause and effect of inner-power-knowledge is to generate a reaction in people (either through their embodied *semanget* or directly on spirits).

At first it could seem that Sakais think that the power lies in the words (Tambiah, 1968). However, the power of inner-power-knowledge to affect *semanget* actually lies in something beyond the words without which the words have no power. During the same time an initiate learns the words of the spell, at night they also receive specific dreams of old people and maidens bearing gifts. These dreams harbour the knowledge that takes form as these images. The knowledge enters the *batin* (*masuk batin*) via *semanget*. The gifts given within the dream-space, as well as the spells learnt within the physical space from the teacher, should be kept as secret knowledge in the *batin*. Through the channel of *semanget* these dreams cause a transformation in the initiate's *batin* which comes to harbour both the words taught by the medicine (wo)man in the physical dimension and the actual 'power to affect' given by his/her spirit-teacher. The words and the power they are supposed to generate become one in the initiate's *batin* as knowledge. When the spells are cast they must be recited with accuracy. The words bring forth the particular 'knowledge to affect' that they are associated with.

So what is this 'power to affect' that the initiate embodies when learning *ilmu*? By way of an answer to this difficult question, I would like to provide an example of inner-power-knowledge, which some Sakai individuals embody. This simple knowledge is called *ago gilo* or 'the mad basket'. *Ago Gilo* is performed for fun. The basic idea of this performance is for the magician to call a spirit to possess a basket, which then starts to move out of its own accord in the hands of the participants who hold it. The following is a transcription from my own video recording.

The Performing Mad Basket

An ordinary Sakai basket (*ago*) was turned bottom up and a stick was passed through the rattan gaps to jut out on both sides. Participants pulled a shirt over the basket and half a coconut shell was attached to it and wrapped with a piece of white cloth resembling a headdress. The basket was thus given an image of the upper half of a human body, the coconut shell and the stick resembling a head and shoulders. The magician first mumbled a spell over the basket. Two people volunteered to hold the basket. They each sat cross-legged facing each other and held the basket up by its rim. They began to gently rock the basket left and right. The magician sat behind one of the participants who was swaying it. He recited a spell, which repeatedly exhorted the Spirit-of-the-Basket to come down and enter it. As he was reciting the spell his right hand swayed a white cloth in tempo with the participants rocking motion. As the swaying motions carried on, the participants began to feel the basket getting heavier. The basket began to rock faster and violently in their hands. After a while the coconut

> shell and the stick were jilted out of place and flung to different corners of the room. The magician unperturbed carried on reciting the spell and swayed the cloth keeping to the original tempo. The only way to stop the basket was for either one of the participants to let go. At exactly the very same moment when either one of the participants did let go, the magician immediately stopped swaying the cloth.

In local conceptualization, the magician embodies the knowledge to call the spirit of the basket to enter it and take over its movements. As the magician recites the spell, the words are understood to enter the spirit-dimension, and the spirit responds by entering the basket. The basket takes on a will of its own, and therefore it is an aberrant (mad) basket personified through its human dress for further affect. Sakais are aware that, although the basket gets heavier 'as though it were filled with water' (possessed by the spirit), it only does so when two people hold it from either side. We might be inclined to read in the magician's spell recitation the power of suggestion, not that dissimilar for example, to the hypnotist's 'hand and arm suggestion' technique. The difference is that whereas the hypnotist speaks directly to the hypnotized unconsciousness, the magician's words are directed to something other conceptually existing in another dimensional space, but suggestively rebounding back to the participants. There is also something else working here. Merleau-Ponty (1962: 185) tells us that the spoken word is a gesture, and like all gestures it comes about through the reciprocity of intentions with other people's gestures. He writes 'it is as if the other person's intentions inhabit my body and mine his.' The words gains embodied meaning in the sensuousness of the interactive context.

In order to have obtained knowledge of inducing the basket 'to go mad', the magician memorized the spell from his teacher. At the same time he had dreams of the knowledge entering his *batin* (inner). As already mentioned, dreams of this sort are a transformative experience of consciousness. They transform the individual's embodiment into an authority for the particular knowledge that he is learning. When the spell is 'raised' from the *batin*, it 'activates' the magician's embodied knowledge, which in turn creates a sensually authoritative interactive frame for this knowledge to suggestively have an effect on the participants. The words are heard within the authoritative embodied space generated by the magician's presence. Although the participants' rocking movements get faster than that of the magician's, their arms are sensually caught within the sway of his firm and authoritative hand movement. This experience is the 'weight' that they are experiencing as the sense of their hands is dissociated from the sense of the rest of their bodies. When one of the participants lets go, the magician senses (*'aso*) the release and stops swaying the cloth at exactly the same moment. Thus, with only the spell but without the embodied knowledge to create the suggestive frame, the magician would have no affect. The ability to create the suggestive frame with his body was gained through the dream experiences that transformed the magician's *batin* (inner) to accommodate this particular knowledge, making it part of his embodied self, and when put to use, of his authoritative embodied presence to cause the intended affect in this particular mode of social interaction.

For Sakais, knowledge from the spirit-dimension is a transformative experience of the 'insides'. The interactive affect of this knowledge is only possible in a context whereby dream experiences are conceptually and experientially conceived within a dualist framework so that knowledge can be experienced as passing from one dimension into another. There is a processual interplay between an individual's experiences in the non-physical space generated by the altered states of consciousness and the embodied self in the ordinary states of awareness. In this culture of embodiment, dualism as experienced through relationships of the self and other in the altered state of consciousness is a pre-requisite for embodied experiences. The knowledge that one embodies from that dimension forms a constituent of the particular inner-self (*batin*) in interaction with others in the physical dimension. Descartes may have disparaged all phenomenal experiences that one could gain from the altered conscious states but nevertheless the fact remains that at least in the above example, the magician has the embodied ability to make the basket 'go mad'.

Conclusion: The Methodological Commensurability of Dualism and Embodiment

We have seen that the phenomenal problem for Descartes' consciousness was 'how do I know that the person or the table in front of me exists beyond my experience of it'. For Sakai this is not a problem. For them the problem is what to make of my experience of a flimsy image of a person, or a person who, when I turn around, suddenly vanishes in front of me, or of a strange lucid dream I may have. Sakai epistemology does not start with conscious awareness, but the universal predisposition of consciousness to alter from an ordinary state of awareness. If for Descartes the inner eye of consciousness could lead him to a rational understanding of his mind-self on the one hand and the physical world on the other, for Sakais, the inner-eye (*mato-batin*) can lead not only to embodied knowledge but to knowledge from another dimension.

Descartes' philosophy is embedded in his religious worldview (Shapin, 1996, Strathern, 1996). He inherits his dualist distinctions from Christian monotheism and the scholasticism of the era preceding him (see Russell, 1946; Strathern, 1996: 161). His ontology is God, mind as soul and matter/body (Wiessman, 1996), a triad that is reminiscent of the Sakai's own spirit/*semanget*/physical body triad. Like Descartes' triad that collapses into a mind/body dualism of substance, Sakais' triad collapse into a spirit/physical dimension. However, Descartes' relationship to God is not one based on religious experience. His ability to know the world is credited to what early twentieth-century anthropology called the 'sacred' but one rationally conceived. His aims were to argue rationally for the fundamental tenets of his religious monotheistic culture and the development of an open rational scientific-culture (Strathern, 1996). For this aim Descartes essentialized reason to such a degree that he conceptually extracted it from common human experience and common-sense. In turn he laid the foundations of a modern culture in which reason itself became reified and anything that fell short from this cultural reification was considered, at best, second rate and, at worst, aberrant (Saas, 1992).

Strathern points out that Descartes conflates the Christian concept of soul with mind (Strathern, 1996: 41). In doing so, he conflates the sense of 'self', which religiously is conceived as detachable from the body, with embodied thought processes that are experienced as part of the conscious body in the ordinary state of consciousness. Hence in Descartes' philosophy, his mind stands aloof from the body, like the 'soul' before his numinous (which in most religious experience is experienced through a different conscious state), but yet has all the properties of embodied thought within the ordinary states of consciousness. Descartes' soul/mind relationship with his numinous was one based on reason, his ultimate 'religious experience'.

As a religious scientist Descartes does share a dualist structure of thought with most peoples of different societies whose knowledge is based on asserting the existence of deities/spirits/souls on the one hand and matter and physical bodies on the other. It therefore would seem anthropologically strange to deny dualism in ethnography when most peoples' views of themselves in the world are based on some form of a dualist conception (see also Finkler, 1994: 181; Lambek, 1998: 106; Halliburton, 2002). As Lambek (1998) points out, it seems to be a universal fact that people seem to need more than one term to describe their experiences of consciousness. The fact that human consciousness can dissociate itself from itself, generate dream worlds, self-reflect and create dualistic domains for metaphorical thinking, is suggestive of a dualistic experience to consciousness. Descartes' philosophical dualism may be understood as one expression of the human conscious condition of experiencing a fundamentally physical and a non-physical duality. However, it may not be so much the fact that humans experience their world within a dualistic framework that is important but what people do with their phenomenal dualisms once they are consciously generated. It might be the case that rather than deny dualistic thinking, we should explore what is the nature of people's existential dualisms in their own semantic fields rather than reducing them to a variant of the Cartesian type. I suggest that these dualistic frames become the pre-conditional frames for the embodiment of knowledge, experience and transformation of consciousness.

Reducing embodiment to a mere critique of dualism theoretically locks it within a dualist framework (see Lambek, 1998). It might be plausible to suggest that dualism might be considered to be the very precondition of embodiment. This would allow us to see the two working on each other through an experiential dialectical process. It also suggests that dualist thought and embodiment are methodologically commensurable.

Notes

1. Animals also have *semanget*, albeit less so than humans.
2. Likewise the shaman's travel in the spirit dimension is an experience in a space that is both external and internal to him or her.

Bibliography

Astuti, R. (2001), 'Are We All Natural Dualists? A Cognitive Development Approach', *Journal of the Royal Anthropological Institute*, 7: 429–47.

Benjamin, G. (1979), 'Indigenous Religious Systems Of the Malay Peninsula', in A. L. Becker and A. A. Yengoyan (eds), *The Imagination of Reality: Essays in Southeast Asian Coherence Systems*, Norwood, NJ: Ablex Publishing Corp.

Cohen, A. P. and Rapport, N. (eds) (1995), *Questions of Consciousness*. New York: Routledge.

Csordas, T. (1990), 'Embodiment as a Paradigm for Anthropology', *Ethos*, 18: 5–47.

Descartes, R. (1996), 'Discourse on Method of Rightly Conducting the Reason and Seeking for Truth in the Sciences', in D. Weissman (ed.), *Discourse on Method and Meditations on First Philosophy, Rene Descartes*, New Haven and London: Yale University Press.

—— (1996), 'Meditations on First Philosophy', in D. Weissman (ed.), *Discourse on Method and Meditations on First Philosophy, Rene Descartes*, New Haven and London: Yale University Press.

Fabian, J. (2002), *Time and the Other: How Anthropology makes its Object*, New York: Columbia University Press.

Fernandez, J. W. (1986), *Persuasions and Performances: The Play of Tropes in Culture*, Bloomington, IN: Indiana University Press.

Finkler, K. (1994), 'Sacred Healing and Biomedicine Compared', *Medical Anthropology Quarterly*, 8: 178–97.

Foucault, M. (1965), *Madness and Civilization, A History of Insanity in the Age of Reason*, New York: Vintage Books.

Halliburton, M. (2002), 'Rethinking Anthropological Studies of the Body: Manas and Bodham in Kerala', *American Anthropologist*, 104: 1123–34.

Hallowell, A. I. (1955), *Culture and Experience*, Philadelphia, PA: University of Pennsylvania Press.

Howell, S. (1984), *Society and Cosmos: Chewong of Peninsular Malaysia*, Chicago: University of Chicago Press.

Lambek, M. (1998), 'Body and Mind in Mind, Body and Mind in Body: Some Anthropological Interventions in a Long Conversation' in M. Lambek and A. Strathern (eds), *Bodies and Persons, Comparative Perspectives From Africa and Melanesia*, Cambridge: Cambridge University Press.

Merleau-Ponty, M. (1962), *Phenomenology of Perception*, Routledge: London and New York.

Midgely, M. (2001), *Science and Poetry*, London: Routledge.

Noll, R. (1983), 'Shamanism and Schizophrenia: A State Specific Approach to The Schizophrenic Metaphor', *American Ethnologist*, 10: 443–59.

Pavel, T. (1996), 'Literature and the Arts', in D. Weissman (ed.), *Discourse on Method and Meditations on First Philosophy, Rene Descartes*, New Haven and London: Yale University Press.

Russell, B. (1946), *The History of Western Philosophy and its Connection with Political and Social Circumstances from Earliest Times to Present*, London: George Allen & Unwin.

Saas, L. (1992), *Madness and Modernism: Insanity in the light of Modern Art Literature and Thought*, New York: Basic Books.

Scheper-Hughes, N. and Lock, M. M. (1997), 'The Mindful Body: A Prolegomenon to the Future Work of Medical Anthropology', in P. J. Brown (ed.), *Understanding and Applying Medical Anthropology*, Toronto: Mayfield Publishing Company.

Shapin, S. (1996), *The Scientific Revolution*, Chicago: University of Chicago Press.

Skeat, W. W. (1965), *Malay Magic: An Introduction to the Folklore and Popular Religion of the Malay Peninsular*, London: Frank Cass.

Strathern, A. J. (1996), *Body Thoughts*, Ann Arbor, MI: University of Michigan Press.

Tambiah, J. S. (1968), 'The Magical Power of Words' *Man*, 3(2): 175–208.

Toulmin, S. (1996), 'Descartes in his Time', in D. Weissman (ed.), *Discourse on Method and Meditations on First Philosophy, Rene Descartes*, New Haven and London: Yale University Press.

Weissman, D. (1996), 'Metaphysics', in D. Weissman (ed.), *Discourse on Method and Meditations on First Philosophy, Rene Descartes*, New Haven and London: Yale University Press.

Wikan, U. (1989), 'Managing The Heart to Brighten the Face and Soul: Emotiona in Balinese Morality and Health Care', *American Ethnologist*, 16(2): 294–312.

Yousuf, G. (1991), 'Semanget: A Note on the Traditional Malay Concept of the Soul', *Kajian Malaysia*, 9: 1–12.

Index